The Essence of Human Existence:

An Exploration of the Interdependence Concept

Sam Alfa, PhD.

Published in the United States of America.

Table of Contents

About the Book - The Essence of Human Existence

The concept of dependence, independence and interdependence has been taken to the extreme to the point that it affects human existence. The extremity has redefined the meaning of human existence, and thereby denied humanity the real essence. In this book, there is an exploration of what those terms (independence, dependence, and interdependence) really mean, and a diagnosis of their impact on human existence.

About the Author - Sam Alfa, PhD.

SAMUEL A. ALFA, PHD. is a trained philosopher, educationist, accountant, and administrator, whose interest has been in what makes living a meaningful experience for every human being. Since his exposure to the phenomenological concept in Philosophy, he has developed interest in researching alternative approaches to the abstract methodological ways of philosophy. He has always observed that there is more to human beings than just the outward observances. Driving this idea further, Dr. Alfa decided to explore the connectedness between human beings and how that connection brings out the essence of human existence. The inherent nature of human beings draws out the commonality that exists among human beings; and that commonality is more evident in the interrelatedness. In the interrelatedness, the commonality is exposed and finds its meaning. In unrelatedness, the commonality is completely lost, and the essence of human existence is thereby lost as well.

Dr. Alfa decided to explore the concept of the things that human beings cherish a lot and attempted to add color to it by turning it into a more meaningful experience for human existence. This is the concept of 'Independence'. Does this cherished concept truly exist? What does it really mean? What values does it add to human existence? Do the concepts of interdependence and dependence bring out the true meaning of human existence? These are some of the questions that are explored in the pages of this book.

Preface

If there is anything that the study of Philosophy, especially, Phenomenology, as a subject, has done for me as a human being, it is that it has helped me to embrace the true meaning of existence. Philosophy has empowered me to not take anything at face-value, especially, things that require some analytical and logical evaluation. It has also given me greater understanding of the interactions that exist between the phenomenal and the noumenal. It has led me to give thoughts to that which unifies hu- mans and the pursuit of a life that would be better than what we live presently. In the world that is full of discord and rancor, Philosophy has engaged me in the search and pursuit of that which can lead to the common goal of our existence that can be more tolerable, or even better yet, be more conducive. While a utopian existence might be out of reach by humans, it is very pertinent to discover alternatives that can be more attainable. These thoughts have led me to the discovery exercise as displayed in these few pages that attempt to examine human existence and what could give or have given it the well-deserved meaning.

The thought process that set me to embark on this heuristic journey, which is on-going, came from a simple analysis of what '*independence*' really means. The conclusion that seemed to be on the horizon of my thought did not augur well with me because it did not give a great meaning and contribution to existence. What led me to the dissatisfying conclusion was the fact humans find it so difficult to let go of whatever they have ever had a subject, especially, in servitude. The desire to hold on to power reigns in the mind of the ordinary human. If one thinks that freedom is granted to another person, then why continue to struggle with power over the person, and try to laud it over them? This nagging question made the concept of *independence* to be very distasteful to me. This dissatisfaction caused me to focus on the simple '*dependence*' since it emanates from the compound word '*in-dependence*'. '*Dependence*', for me, even produced a more disturbing outcome in my thought because I did not see it as contributing anything meaningful to existence if it stands on its own. On its own without reference to anything else, *dependence* was more burdensome and boring. Out of my curiosity and inquisitiveness, I decided to press a little harder, and that landed me in the world of the compound word: '*inter-dependence*'. After much thought processing, I found a niche for the probable remedy.

I was at a wedding (October 2021) where I was called to give a word of advice to the newlywed couple. It was an impromptu invitation to say something and give a word of advice. I stood up and said the following:

In this marriage, eschew living independently of each other be- cause it is a myth, that means that there is nothing like that; eschew living dependently, because it is burdensome; but live interdependently because that is the way to succeed. As two in- dependent people, you must live dependently by relying on each other interdependently if you must succeed on this journey of marriage. Do not live independently of each other because you need each other, otherwise this marriage will have a sour taste. Do not live dependently because you will each become a burden and give the marriage a bitter taste. Live interdependently be- cause that is what life and marriage consist of, and you will enjoy a sweet taste.

The moment I sat down, it dawned on me that I had examined these concepts for a while, and it was time to pen these thoughts on paper. I turned to my wife, Dr. Fatima Alfa, and shared the thoughts. I am indebted to her for this project, especially, for serving as my scribe when driving on the road. I am also indebted to Sime- on Alfa, my brother, who counseled on the better way to approach the write-up.

My aim is that this will turn into a project for humanity, globally, especially, as we have deviated from this path of human existence. We must return to what will give sanctity to human existence and humanity, and the project starts now. This will be an all-encompassing project that requires everyone's attention and involvement, as we work to arrest the devastating reality of human disengagement that is looming on the horizon or that has started eroding the meaning of human existence

Introduction

In the world, there is a great quest by every human being, and it is the pursuit of independence. The quest for independence has been one of the major challenges of humanity from the time the world began. For one reason or the other, humans feel trapped by something or feel like they are controlled by somebody or something, and hence want freedom. A human being craves independence from everything without really knowing what the quest entails. We want independence because we feel that we are trapped by the fellow human whom we consider to be above us, suppressing us, and oppressing us. We have a feeling that there is something better that we are being deprived of, hence we want to be released so that we can really explore that thing for ourselves. Human beings are never satisfied with any explanation from another person because our thinking is just about getting free from the illusion of the prison of life, and most of the time, it is self-imprisonment. As a matter of fact, we even con- sider the explanation as a way of tying us down to that prison. At this level, human beings feel that they are being disadvantaged or that the person who appears to be helpful is also one who is attempting to deprive us. While *independence seems* to be great, mark the word, '*seems*', it carries some dire consequences beyond human control, but the inquisitive hu- man mind so capitalizes on the quest for in- dependence that he/she continues to pester mankind, thereby making us to think that we will discover the something that others are depriving us of. The mind will continue to engage our emotions leading us to the path of irrational actions instead of rational decision based on informed knowledge. To be truly independent and self-reliant, it takes sacrifice and determination.

Why does mankind spend so much time in the pursuit of independence? Some of the reasons are:

1. The '***Feel-good***' syndrome – It just feels good to be independent. The quest for independence, based on this syndrome, is too shallow as feeling good is always a temporary thing and almost usually, is momentary. After the feeling dissipates, then what?
2. The '***I want to be like him/her***' syn- drome – This is the desire to be like other people either from their appearance or attitude. For some it is borne out of envy or jealousy or unhealthy competition. We fail to realize that our situations are different; hence it is not one size fits all for every human being. No one has an idea of what the other person is contending with, hence the desire to be like the other person is based on emotions which vanish quickly. The competitiveness in a human being leads to the quest for independence,

and it is mostly competing against the unknown.

1. ***Lack of Trust*** syndrome - Sometimes, the quest for independence is because of our suspicion of the people we deal with; but we fail to realize that it comes with consequences that are very costly. The fear of manipulation and ulterior motives pervade our reason for seeking independence because we are not quite sure of what the other person is thinking. You can imagine the situation of being cheat- ed by the other person. The resulting fact is that you will try to prepare against the unknown. The fear of being cheated can lead one to becoming a cheat. The quest for independence could eventually turn one into a n imaginary fighter.

The question must be asked: What is *independence*? Can independence be described as freedom from other people, isolation from others, being an island, etc.? What then do we make of the common saying: ***No one is an Island?*** If that is the case, then how does in- dependence become such a hot pursuit by the young people, especially, where their parents no longer matter or count in the affairs of their life? The lifestyle of the young adults is lived apart from their parents and family members, resulting in a life of isolation, alienation, and separation.

The issue of independence is not only with human

beings. The same is applicable to almost everything. Take for instance, there is no country that has attained her independence that is completely on her own. No country trading by herself will be measured on the stage of the world market as having a strong economy. There must be the level of interaction and ex- change. There remains some level of allegiance to the 'former master'. If independence is truly independence, then all such ties must seize, but that is never the case. The former master is always seeking for ways to exercise some level of control and ownership, while the former servant is looking for continued assistance from the former master. In some cases, when one does what the other is not pleased with, there could be consequences, but in the real sense of independence, nothing like that should exist. Even when the so-called independence is supposedly granted, there can never recognition by the former master that the former servant is an equal partner on the same stage. For independence to be truly independence, especially, on a global level, there is must be a change in status to equality for fairness on the playing field. That is not the case. It is the lack of in- dependence in national and internal relations that has led to neo-colonialism, but in the real sense it is not neo-colonialism because there has never been complete freedom or independence. It was just imaginary, not real. The innate desire in human to control or to be in control denies the essence of independence; and the fear to stand on oneself, especially, after being subservient for a long time, is a strong challenge to independence.

Many people or nations do not stop to think of what '*independence*' means. Many people do not even take a step further to consider whether a life of independence is how humans are formed or configured to live. Many do not even consider what the consequences of independence could be. In my reflective thinking of this thing called '***independence***' which many people, nations, children, and so on, aspire for, I had to ask some serious and tough questions including but not limited to the following:

What does independence mean? Is independence real?

Were humans created as independent beings? If we understand the core configuration of a human being, is independence what is designed for humans?

What are the benefits of independence?

Are there characteristics of independence that we can identify with? What are those characteristics?

What does independence mean in terms of a nation? Can any nation be truly independent?

What are the alternatives to independence, and how do they connect, if any?

Does independence mean freedom?

How does 'in-dependence', which is: 'not dependent', mean 'freedom'?

In my search for answers to the questions above, I started to reflect on the fact that there are serious challenges and consequences of independence which basically short-change the human being and challenge the well-being of every human. I also discovered that there are great benefits of independence. In my quest for alternatives to independence, I considered *'dependence'* and '*interdependence*'. In the following pages, I will explore each of the alternatives and their ramifications to the human existence.

Chapter 1 - Dependence

Definition and Description

According to the Merriam-Webster dictionary, ***dependence*** means *the state of needing something or someone else for support, help*. It is the *state of relying on or being controlled by someone or something else.* It is also reliance on someone or something for support. ***Dependence*** is a state where someone is heavily reliant on someone or something else.

Dependence is a sense of incompleteness without something else. Its completeness is in the addition of something else on which it is dependent. It is the concept of lacking full realization or full potential without the other piece or pieces to complement or supplement it. It is a weak position; hence the power of sustenance is absent until the other pieces are present. Dependence connotes the idea of reliance on a power source for existence or provide the needed vitality. It is inability to act due to powerlessness. It is also the notion of imperfection, insufficiency, inadequacy, lack of competence, helplessness, powerlessness, fragility, insubstantiality, and vulnerability.

Attributes of Dependence

TO GET A FULLER UNDERSTANDING of the characteristics of 'dependence', it would be necessary to examine some of its attributes. Some of the attributes of dependence are:

Helplessness - inability to defend oneself or to act effectively. It means being unable to help oneself; weak or dependent: a ***helpless*** invalid; deprived of strength or power; ***powerless***; in- capacitated: the feeling or state of being unable to do anything to help oneself or anyone else: Doing something for others can stem the feelings of ***helplessness***. It is lacking protection or support.

Weakness - the state or condition of lacking strength. ***Weakness*** is defined as a lack of strength or a character flaw; or something that one cannot resist; the state or quality of being weak. *Free Dictionary* defines weakness as - 1. the state or quality of being weak; lack of strength, firmness, vigor, or the like; feebleness; 2. an inadequate or defective quality.

Defenselessness – the quality of being unable to protect oneself from attack; the fact of having no protection.

Vulnerability - Vulnerability is a weakness in a system or device that can be exploited to allow unauthorized access, elevation of privileges or denial of service. *Vulnerability* refers to the inability to withstand the effects of a hostile environment.

Subservience - willingness to obey others un- questioningly; the condition of being less important than something or somebody else.

Subordination - placement in a lower class, rank, or position; the act of giving someone or something less importance or power.

Fragility – This characteristic of *dependence* relates to the idea of not being able to maintain oneself, and always being a breakable position and hence needing the support to remain in- tact.

All the attributes/characteristics of ***dependence*** described above can easily be related to human beings. To be dependent on others for one's existence, calls for some serious evaluation as to the value that one brings into life and living. One can imagine the type of life that one would be living in this state of life – *dependence*. We will examine some of the pitfalls of *dependence*.

Pitfalls of Dependence

BASED ON THE FEW CHARACTERISTICS or attributes, we will examine some of the drawbacks, perils, dangers, or downfalls of the life of *dependence*. The concept of *Dependence* can lead to the following feelings:

Depression – the fact that one is dependent on the other can lead to depression, especially, when things do not work out according to one's expectations, which is always the case when you are at another person 's mercy. If the expectations are not forthcoming, it can lead one to the condition of depression. In the words of Atticus poetry "Depression is being colorblind and constantly told how colorful the world is." The person dependent is colorblind but because he is unable to determine the color, he is being told how colorful the world is when it is not colorful. It is a depressed life state when one's expression of life is dependent on another person's expression. Depression is the loss of control of one's own emotions and emotional stability. It is a state of unstable psyche there- by leading to unhappiness. When dependent, your expectations would be, more likely than not, cut short most of the time. The lack of the control of one's own destiny can lead to a state of depression.

Agitation – People who are always under the control of another person agitate for some sort of freedom which is not always achievable. One will always protest the purported boss or the one on whom one is dependent. Someone said: *"To be a human being means to possess a feeling of inferiority which constantly presses towards its own conquest. The greater the feeling of inferiority that has been experienced, the more powerful is the urge for conquest and the more violent the emotional agitation."* The feeling of unsatisfaction due to deprivation by the one controlling always creates the spirit of agitation out of the desire to crawl out of the state of dependence.

Anger – When one is not in control of one's daily activities, the resulting outcome is anger. When one is not at liberty to do whatever he/ she likes, anger will be displayed out of frustration as he/she will not be able to do anything about it. Deprivation of one's desires causes anger.

Anxiety – What else can one do but to live in perpetual anxiety of what comes next from the one on whom one is dependent? Anxiety is the feeling of worry, nervousness, or unease, out of uncertainty outcomes. The person who is dependent has no control of the outcome of anything because he/she does not participate in the decision-making process since his/her voice is swallowed up. The voicelessness of the dependent person leads to anxiety. It leads to the nervousness of what the impact of the decision would be on him/her. A life of uncertainty is a life of anxiety, worrisome, nervous- ness.

If a life of dependence hands one some pitfalls, by no means exhaustive, described above, then, there is much to be desired instead of '*dependence*'. The alternative to '*dependence*' that might seem logical would be '*independence*' which literally means, '*not dependent*'. We will examine and explore '*independence*' in the next few pages to determine if that can be a viable alternative to a life of '*dependence*' in human existence.

Chapter 2 - Independence

Independence: Definition and Description

Obviously, the word *'independence'* shares something in common with the word *'dependence'*. However, the prefix "*in*" causes the meaning of the two words to be opposites. I*ndependence* (independent) is described as not being influenced or controlled by others in matters of opinion, conduct, thinking, acting for oneself; not subject to another's authority or jurisdiction, not relying on another or others for aid or support; refusing to be under obligation to others (Woolf, 2002).

Independence can be defined as the recognition of the individual as a subject (not 'object'), a person; a contributing human being recognized; worthy of making a difference.

We have also read about 'independent thinking'. Independent thinking is the thinking that takes everyone else **out of** the equation. It is a sole thinking without seeking counsel and ad- vice. Such approach or attitude of not seeking others' counsel could be for fear of listening to contrary views or opinions that would be at variance/conflict with the views of the independent thinkers. It prevents one of the opportunities to research and learn about other perspectives which always contribute or lead to a properly informed position on issues. Independent thinking limits one to his/her own and only perspective, and outlook on life. Independent thinking does not entertain advice because it is considered by the independent thinker as a sign of weakness; the idea that they are subject to, or influenced by, another person – *the concept of dependence.* The independent thinker does not consider the fact that failure to consult with others could be a deterrent to one's progress in life because there would be room for lots of unwarranted mistakes that could have been avoided if advice was sought, or if there were varying perspectives that result in a more productive conclusion, based on the cohesiveness and the fusion of the varying perspectives.

An Analysis of Independence

GENERALLY, PEOPLE PREFER to maintain their in- dependence and continue to live in their own world rather than change. While it is great to live in one's own world, it needs to happen in a way that does not lead to **isolation,** which connotes the idea of being alone from other people. According to Dr. G. Allen Power, MD, FACP, in ***Dementia Beyond Disease***[1]: *Enhancing Well-Being*, "Isolation is potentially dangerous because it separates us from a network of support, should difficulties occur" (Power, 2014). The dangers loom is the fact that those who aspire and proclaim independence take

1. *http://www.alpower.net/alpower_Dementia_Beyond_Disease.htm*

that ideology to the extreme by desiring to be free from all other as if that affirms their existence. They equate independence to freedom. Dr. Power further explains that: "No one is truly independent, but because this is held up as an ideal (and dependence is seen as bad and burdensome), people are unwilling to ask for assistance and support when they need it. And since they are unable to do for themselves, they be- come increasingly isolated in their homes, even from their close neighbors." (Power, 2014). Isolation soon leads to **loneliness**, and loneliness has social, economic, mental, psycho- logical, sociological, and medical challenges. Loneliness can lead to loss of social interaction which further detrimentally/negatively affects the physical, psychological, emotional, cognitive, spiritual, and health wellbeing. Those who are isolated generally do not have the concept of seeking for assistance any longer because they have gone beyond the level of seeking for one. According to Sophia Dembling, (2014), isolation/loneliness "can also mean you have no one to help you when you need it—whether it is physical assistance or emotional support." It is very important that we distinguish between ***isolation and solitude***. Isolation is almost like a disease that one no longer has control over even though it starts out as having control be- cause of the quest for independence. Solitude, on the other hand, is a choice that one makes, and it is temporary. Solitude is temporary because it is for a purpose of achieving something, either working on a project, meditation, or for the sake of avoidance of distraction, or even a quest for self-discovery on a spiritual journey. Solitude, therefore, does not connote any negative intent as does isolation which is a deliberate effort to be set apart from other people. In making a distinction between solitude and isolation, *Sophia Dembling*, **Introvert's Way: Living a Quiet Life in a Noisy[2] World[3],** (2014) states that solitude and isolation are not the same things, and that "Solitude can be a very positive thing and it implies choice. Isolation happens when the solitude is no longer a choice."

Independence and Individuate

TO INDIVIDUATE IS TO be singled out or be treated as distinguished from like beings. Sometimes, independence quest by the young people is also referred to as 'individuate' from parents. The teenage years are sometimes among the most agonizing seasons in life—for both parent and child. In the adolescent's quest to "individuate" from his/her parents, they openly reject parental values and rebel against their rules. The adolescents are suspicious that their parents' purposes are merely to make them miser- able. Though they eventually come to agree on those matters, the fact remains that by that time, in their relationship with their parents, the relationship would have been riddled with tension. Parents undoubtedly lament the refusal of their

2. **http://sophiadembling.com/books.html**

3. **http://sophiadembling.com/books.html**

children to heed the wisdom of their instructions, knowing that those instructions would, later in life, spare them unnecessary emotional and physical pains. The idea of *independence,* in this case, largely encourages *disintegration*, especially, in terms of family members who crave for *independence* because they feel that they 'have arrived' of age. This is a major characteristic of 18-year-old young people. The fact remains that many of them act on the emotion- al stage and shallow concept.

In interaction with many of the teenagers who aspire, crave, and pursue *independence*, it is common knowledge that many of them do not have the slightest idea of what they crave for or desire to attain. For many of them, it is a status symbol, especially, when they see their age-mates who tell them stories of their self-ac- claimed independence.

For some, the crave for *independence* is bowing to peer pressure. They just want to be like their fellow teenagers, but they have no idea what it entails. They refuse to know that even identical twins have different perspectives for lifestyle, and hence what is good for the goose might not be good for the gander. They do not care to learn about the family background or the ways the families of their fellow teenagers operate. They have no idea of the financial strength of those families of their fellow teen- agers. They have no idea what led to the decision. They have no idea of the arrangement between the family members, and so on.

The ones who have the mentality to individuate take laws into their own hands as if they are in full view of what the future holds for them. They do not want to learn from those who have gone through these stages in life, especially, their parents. For many of them, they rather experience it themselves than learn from the wisdom of those who have already experienced it. Conventional wisdom demands that no human should make the mistakes of those ahead of them, regardless of what dispensation.

Some young people even accuse their parents of being outdated in their outlook on life, especially, as they claim that the young people are in the jet age as opposed to their parents who belong to another dispensation that is not so much enlightened. They consider their parents as being old-fashioned, but they fail to realize that matters of life resurface and never run out of fashion. The ethics of living remain the same regardless of generations and generational gaps. For instance, failure is failure; success is success; and disrespect is disrespect, no matter which generation. Truth is truth. Falsehood is falsehood. Evil is evil. Good is good. The facts of life and consequences for behaviors are ageless, dispensation-less, timeless, and generational-less. All that we can do is to interpret them differently, but the consequences are the same. These point to the fact that it is a myth to aspire for this thing called freedom or independence without a full understanding of what that means. Many are driven by the desire without thoroughly evaluating what it entails. Many of the young people even move themselves out of their parents' home, and when they realize that their experiences are at variance with their expectations, at that point some of them find it more difficult to own up to their mistakes and misjudgment and return to their parents and family. The ego that mandated them to depart prematurely will take a better part of them such that they will not want to return. For the most part, many of these teenagers end up in un- bearable situations and even death.

Independence and Self-disclosure

SELF-DISCLOSURE is one thing/exercise that is so hard for human beings to embark on. Why? It is a stage of transparency and vulnerability. It is exposing one to criticisms and ridicule. Self-disclosure is not a welcome concept for many people. The drive for this lack of interest in self-disclosure is *independence.* No one wants to be criticized by making himself/her- self open to the world. No one wants to leave himself/herself to unusual ridicule. The fact re- mains that self-disclosure is the choice to go the highway at the expense of many innuendos and unhealthy and erroneous assumptions by other people about one's character. It is easier and more honorable to let the whole world know what you have done than to wait for the world to discover. It is easier to let the whole world know who one is before the world embarks on the research to discover who one is, especially, when there is any dark side. The damage done at the point of the world dis- covering about one is much more than that of self-disclosure. Self-exposure is coming out clean rather than waiting for community-expo- sure. When you have disclosed yourself with- out any more secrets, then people will not have much evil to discuss about you. By the time people begin to share the story, there will be nothing else because it has all been exposed by yourself. Self-exposure is self-exoneration. Self-disclosure is healing. Self-exposure eliminates the fear of being discussed with rumors and character assassination. Self-disclosure is self-liberating. It is freedom from emotion- al, psychological, and sociological trauma and guilt. Self-disclosure is openness and transparency which could have serious repercussions; hence it is not common with people who want to affirm their position and claim *independence*. Who wants to be exposed to be ridiculed? **Self-disclosure is self-openness publicly**. For the person who claims independence, there can be no self-exposure, self-disclosure, and self-openness because he/she does not owe anyone such obligation.

Self-disclosure is not the same as **self-condemnation**. Self-condemnation connotes the idea that one is not good enough or even fit to be in the company of any group. Many times, self-condemnation leads to self-pity which has adverse effects on an individual who has such feelings. Self-condemnation is always a result of being boxed in by one's own failure to be open and to interact. One who self-discloses can overcome elements of self-condemnation because such a person will have the boldness to face the public. The person will not be looking over his/her shoulder for fear of being stabbed. The person who does not self-disclose, be- cause of the claim to independence and one's own life, will more likely tend towards self-condemnation than the one who self-discloses. It is noteworthy that self-condemnation can also lead one to the pursuit of *independence,* where he/she can be isolated.

Independence and Collaboration

ACTING ALONE, *independence*, leads to lots of unknown mistakes because everything is seen in and through the eye of the one and only person. It does not promote collaboration with others. *Independence* does not recognize other people's opinions; and does not encourage collaboration, because independence is perceived as a way of self-discovery, and as the quest for self-identity. The quest for *independence* could be a self-inhibiting act to limit one from reaching one's potential in life, a difficult thing to achieve without collaboration. The result could be suicide, drug abuse, alcoholism, and other vices.

Independence promotes silos which is detrimental to collaboration and synergetic approach to resolution, especially, in the marketplace or in a community of people. Where everybody wants to stay to themselves in a community, there can be no synergy and hence strength diminishes. **Collaboration**, on the other hand, builds synergy which fosters strength to accomplish because of the common focus and singleness of purpose of the team. The idea of oneness of a team through collaboration is lost where *independence* is the watchword.

Attributes of Independence

INDEPENDENT PEOPLE do certain things that others do not. In this section there will be a discussion of some of the common attributes of some of the independent persons. (*A disclaimer that these are not all common with everyone who imbibes the spirit independence, but they could be common traits traceable to such group of people. Also, this list does not exhaust all the attributes either*).

Take Risks

INDEPENDENT PEOPLE strive to better themselves and their life. Making a better life means that they take more risks than others because they are not afraid of the consequences. Many of them do whatever it takes to face the challenges that confront them. Independent people do not want to follow the same path as everyone else. They want to set themselves apart as they work toward their dreams and goals. Many times, because the goals and dreams are not guided for fear of being perceived as being de- pendent and weak, those goals and dreams become unachievable, or they go through un- warranted sufferings before arriving at the desirable goals and dreams.

Develop Habits

THOSE WHO ARE HIGHLY independent know that the things they do become a habit which build their confidence. This is characteristic of those who have a great background with parents who instill some discipline and teachings in their upbringing. They always are developing new habits because they know that those habits will give them confidence. Likewise, there are those who develop habits that will lead to their demise because they were not guided in the development of those habits. That occurs be- cause of the rush to become independent just for the sake of being one.

Open Mindedness

SOME OF THE INDEPENDENT people view approach to life with an open mind because they realize that in life changes and differences are inevitable.

The open-mindedness to these changes and differences helps them to grow and move forward. Openness of mind comes with being receptive to other people's perspectives and opportunities, but more likely than not, those who are rebellious in their pursuit of independence are closed to new perspectives. The challenge is that when such is contracted under rebellious atmosphere, the made-up minds are unwise and thereby lead to undesired consequences. When an independent person makes up his/her mind, he/she is not easily swayed. He/she listens to the ideas and opinions of others but sticks with his/her decision without hesitation. The reason for not easily following the advice is because such people want to feel strong. Once an independent person has made up the mind, it is hard to convince such one to change it. Their beliefs, values, and desires are the only things they consider when making decisions. They believe in themselves and trust that their instincts are correct no matter what.

Disconnectedness

FOR SOME OF THE INDEPENDENT persons, ceasing to interact with their family and some friends is a way for them to express that they are on their own as independent persons who are in control of their own destiny. Much to the dismay of their loved ones, independent people sometimes go days/weeks/months, and even years without talking to any of their loved ones.

Constant communication is not something they desire in their lives. They will not care what happens to their loved ones. Even if they call later, they appear to be disconnected without showing any sign of connectedness. Communication, which is key in every relationship or relatedness, becomes very insignificant in the eyes of such independent persons.

Self-Motivated

MANY INDEPENDENT PERSONS do not wait to be told of what to do before they start on what needs to be done. When they see anything that needs to be done, they get started. For the serious ones who build such good habits, it helps them to become successful because they are focused. The reason why they do not need to be asked to do anything is because they are observant of their surroundings, and it portrays them as being in control of their own decisions. They do not care to seek counsel for a better and solid foundation for a more desirable out- come.

The Pitfalls/Adverse Effects of Independence

WHILE BEING INDEPENDENT has some great at- tributes, there are some pitfalls and adverse impacts on being independent that need some exploration.

Stress - **Stress** is how we react when we

feel under pressure or threatened. It usually happens when we are in a situation that we do not feel we can manage or control. It is true that everyone can be stressed, but the difference is that when you operate in the environment that you are not alone, you can experience some burden-sharing which goes a long way to address the stress. When there is no one to share such burden, the resultant effect is devastating. Stress is the sign of inability to cope.

Loneliness – Independent people may not admit it, but loneliness is a major characteristic. In their quest for independence, they are always left alone

Depression – Depression is a feeling that is not definable because it deals with moods that affect emotions. It is more describable by those who go through that experience. It is a feeling of being unsure of what one needs while at the same time knowing what one needs. It is a feeling of unattainability. An unknown person describes it this way:

> *Having anxiety and depression is like being scared and tired at the same time. It's the fear of failure but no urge to be productive. It's wanting friends but hating to socialize. It's wanting to be alone but not wanting to be lonely. It's caring about everything, then caring about nothing. It's feeling everything at once, then feeling paralyzingly numb.*
> – Unknown.

This is the situation with the one who exhumes the feeling of independence. It is a deep sense of feeling that is not easily admissible by the person under that experience. It comes with that sense of denial of the reality because of false self-affirmation. While the body wants to survive, the mind wants to die at the same time. It is like feeling a very heavy weight on one, but also finding comfort under that weight because, for some reason, there is that sense of false or pseudo security, all in the desire to display independence. When this depression is not handled very well and dealt with, it can give room for suicidal thoughts to creep in, and with the complete isolation, suicide could become the reality.

Anxiety and Worry – This is another aspect of depression. The feeling of carrying the bur- den of the whole world by one person creates a heightened sense of anxiety and worry. This sense is created by the feeling that one is alone. It is a feeling that everything and wall around one is collapsing. This feeling comes from the sense of the lack of support.

Greed – Some independent people do not feel a sense of responsibility to anyone, and hence they keep everything to themselves rather than living with an openhand. They continue to ac- quire more and more, but for themselves only. It does not matter what the other people are feeling around them.

Distrust – The fear of being controlled leads

to distrust. Generally, the feeling of not able to trust another person is very prevalent in the life of an independent person, hence the desire to stay away from other people. This is also a feeling of doubt about some person. It can even escalate to the point of considering every other person as being dishonest.

Fear of communication - Fear of listening – One of the major elements of human existence is communication which literally means interaction with one another. The chronic independent person fears the art of communication for fear that he/she might be influenced by the discussion.

Fear of obedience – For the chronic independent person, obedience can be a sign of weak- ness or subjection to a higher authority which could be construed as deprivation of one's own liberty or freedom. The failure to obey could then be seen as disloyalty.

Chapter 3 - Interdependence

Definition and Description

In my quest for alternatives or complementary factors to independence, I considered 'dependence' and 'interdependence'. Dependence had challenges that make people and nations run away from it. Then I further considered 'interdependence'. That is where my research found an alternative to 'independence' but not at the expense of independence and dependence. There is still a level of connectedness which addresses the concept of interdependence. That is what dependence and interdependence truly mean.

The prefix 'inter', in the compound word 'inter-dependence', means *in between, within, among, together, in the midst of,* be dependent on, *mutual, reciprocated, or active during an event.* Etymological study of the word 'interdependence' reveals that it is a compound term from two Latin words: 'inter' and '*dependere*.' It simply means dependence among or between two independent things. The dependence on each other or one another is mutual reliance and reciprocal. It is a symbiotic relationship be- tween two or more independent things or hu- man beings such that one is incomplete without the other, not in terms of the physical outlook but in terms of consciousness.

Interdependence is being mutually dependent, or simply being dependent on each other (Woolf, 2002). Two people in a healthy relation- ship are said to be interdependent. In contrast to existing alone, it is a voluntary recognition that "no man is an island," and that we must co-inhabit the space in which we live. It is worth noting that those who are mutually dependent on one another must themselves be independent, that is, able to stand by themselves. If one is not able to stand alone, then there would be challenges with such mutual dependence, and as a matter of fact, it seizes to be mutual. In other words, those who mutually depend on one another can sustain themselves to a certain level before finding that extra support for completeness. Therein lies the independence intertwined with dependence. The most important point to understand, according to Kim (2006) is that 'this debate positions both independence and interdependence as alternative endpoints of maturity... Independence and interdependence are opposites—either the two ends of a continuum, or else mutually exclusive categories" (Kim 2006:146).

Interdependence is collective brainwork which is always better than single brainwork. That is why we need interdependence. It connotes the concept of **collaboration**. Collaboration identifies and recognizes the strengths of the individual and embraces the knowledge and insights of others. Collaboration is key to achieving global impact of anything that we step out to achieve.

Interdependence is working together with a common, shared purpose. Each organization has clarity about its own distinct, unique role and the roles of others, as well as a holistic understanding of how they interact to create cohesion across organizations. It is living within a network of contact and mutual support.

Inherent in interdependence is stewardship which connotes the idea of responsibility for oneself and others. **Stewardship** is taking responsibility for forming working relationships with others to transform well-being across a region, together. It is the practice of encouraging alignment, inclusiveness, resident leadership, and knowledge sharing, marked by a commitment to the regional effort beyond one's own vested interests. There is the drive to carry other people along which is a sense of inclusivity.

The concept of **team** is at variance with independence and dependence, but finds a fitting abode in interdependence. Team promotes co- operation of all and from all. ***'Of all'*** because all are the same in different ways and each has a unique contribution in a collaborative atmosphere. ***'From all'*** because all are integrative part of the whole, and all have the unique contribution to the universal good.

Integration is another aspect of interdependence which takes all to be integrated into one another. The independent is integrated into the others as if it is dependent, but it is interdependence.

Accommodation, as a concept of interdependence, stresses the fact that everybody is accommodated. No categorization, no dis- crimination, and no exclusion. Acceptation and acceptability are core ingredients of interdependence.

Empowerment is an aspect of interdependence because it acknowledges the contribution of others, and that acknowledgement empowers others to contribute because they are valued. This is how interdependence relates to empowerment. The recognition of the unique- ness of the independent individual empowers participation and contribution. Interdependence is Empowerment. In Empowerment:

- No one is an island
- Everyone has something to contribute
- No one is useless
- Everyone is incomplete without the other person(s).

The concept of **Wholeness** and **Complete- ness** resonates with the concept of Interdependence.

The concept of interdependence brings out the

hidden virtue residing in everyone.

One of the major purposes of life and your existence is to have an impact on another per- son or on persons, hence failure to do that is to live a less impactful life and to lack **fulfillment**. Your life and mine are incomplete and lack fulfillment without the ingredient in you or me. The flavor of a great dinner is in the mixing of different spices, so also is the meaningfulness of living. Try using one spice for a meal, and you would notice how stale it is. The beauty of a dress is in the multiplicity and alignment of the various colors.

There is something unique about you and an- other person, and the **uniqueness** in me loses its uniqueness without the uniqueness in you.

Interdependence brings out the concept of **one small world** or the **global village** because no one can determine who carries what is needed to complete the other person(s). The concept of interdependence goes beyond race, color, age, religious affiliation, marriage. This concept of interdependence is **classless.**

It is not **communism.** It is not an ideology. Interdependence broadens the concept of

connectedness and interconnectedness. As a matter of fact, without interdependence, there can be no interconnectedness because if I do not feel that I need anything from you, I will not find the necessity to be connected. How- ever, it starts from the concept of the mindset that we are all human beings and that we need each other. Phenomenologically, what gives me the realization that I am a human being is the self-consciousness that there are other like people who behave like me, look like me, act like me, and display similar characteristics.

As a matter of fact, interdependence brings out the concept of **relatedness**, not 'relationship' as we know it to be. While in relationship, one can deceive another; in relatedness, one would feel that one is deceiving oneself because one will be depriving oneself of the full benefit of what gives one a fulfilled life and the complete- ness that one needs.

The concept of **hatred** would be a far cry as no one would love to hate oneself. Relatedness is more than **acquaintances or friendship** because one could decide not to make any contribution to the relationship, especially, since many people go into relationships for different reasons.

Relatedness is not **fellowship.** It is an agreement to go in the same direction collaboratively and cooperatively.

Interdependence forces **participation** because no idea is non-sensical. One's idea is pertinent and relevant to every discussion.

Interdependence brings out the **humanity** in us, but it is not **humanism.**

Interdependence defines equal and same level

playing field. It goes against the concept of godfatherism and favoritism.

Interdependence is **Enlightenment**. It is **emancipation from tutelage**, from **subordination**, from **subservient** position, from **self-centeredness**.

Interdependence is **liberty from inhumanity.**

Interdependence is the antidote to **racism** and

classism.

INTERDEPENDENCE IS **belongingness.** Everyone belongs to everyone thereby contributing to a mutual and formidable force.

Interdependence thrives under **mutual feeling.**

Interdependence is **engaging/engagement.**

Interdependence is **interaction.**

Interdependence defies **superiority and inferiority** complexes.

Interdependence thrives in a **blameless** society.

Interdependence is foundational on **trust** and

dependability and reliability.

ATTRIBUTES OF INTERDEPENDENCE

There are certain attributes that pervade the definition or description of the word 'interdependence', and validate the terminology, such as: mutualism, team, alliance, and correlation, to mention but a few.

According to the *Merriam-Webster dictionary*, **mutualism** was defined as "mutually beneficial association between different kinds of organ- isms". Mutualism is a key element in the inter- dependence concept analysis because it high- lights the importance that two or more different organisms/entities equally benefit from their relationship. The *Merriam-Webster dictionary* defines **team** as "a group of people who work together" and "marked by **devotion** to teamwork rather than individual achievement". Being a member of a team is essential to the concept of interdependence because it emphasizes the importance of working together to achieve a common goal.

The *Merriam-Webster Dictionary* lists the definition of alliance as "a relationship in which people agree to work together". The attribute of **alliance** is an important aspect of the concept of interdependence because it displays the requirement of a joint agreement to complete a task. **Correlation** is defined by the *Merriam-Webster Dictionary* as "a relation existing between phenomena or things or between mathematical or statistical variables which tend to vary, be associated, or occur together in a way not expected on the basis of chance alone". If not based on chance, then it is based on mutual agreement. Correlation is an asset to the concept of interdependence due to the relationship and relatedness that exist between two items that appear in accord.

The attributes of interdependence: **Mutualism, team, alliance,** and **correlation** help express the need for a mutual relationship to efficiently accomplish a task together. These attributes make the concept of interdependence to be valid.

Where interdependence exists, all these attributes must be present. These attributes are embedded in collaboration which is a much-need- ed attribute in every model. Take for instance, in health, this model case shows how collab- oration of *care* is of great significance in the health care setting. All the attributes of inter- dependence are displayed within this situation in relation to members of the healthcare team. One branch cannot work fully without the others. The patient depends on the nurses to take care of them, the nurses depend on the physicians to create diagnoses and write orders, and the providers depend on the members of the care team to see that their orders are completed. Good teamwork between nursing and medical staff is important for improving patient safety, avoiding medical error, and ensuring patient outcomes and staff well-being. Each component of a health care team is mutually reliant on each other and share reciprocal relationships. Interdependence delivers a great result which is desirable for all, in this case, the patient gets well and becomes healthy. This is the output of an interdependent society; it is well and healthy.

Antecedents and Consequences

THERE ARE CERTAIN ELEMENTS that must be present to effectuate the attributes of interdependence. The elements help to refine the attributes of interdependence, and without them there can be no interdependence. Walker and Avant, (1994) describe them as antecedents which are "events or incidents that must occur or be in place prior to the occurrence of the concept" (p. 167). When dealing with the concept of interdependence, some of the antecedents are risk-taking (faith/trust), acceptance, char- acter, and circumstance.

Risk taking is also referred to as **Faith**. Faith has been defined as the "strong belief or trust in someone or something" (Walker and Avant, 1994, p.18). Faith is shown as one of the most essential aspects of interdependence because faith and trust in others are key for relationships with and dependence on, other persons.

Acceptance of one another and understanding that systems must work together is conveyed as another essential antecedent of interdependence. *Merriam-Webster* Dictionary defines acceptance as "agreeing either expressly or by conduct to the act or offer of another," which is key when working with others.

The next antecedent of interdependence is described as **character**. Character is shown in relation to the qualities that make up how someone thinks or feels. Character also reflects a person's morals and belief systems. "Interdependence stimulates and challenges individual opinion, guides and augments initiative, and demands individual responsibility and accountability, attributes strongly associated with professional ... autonomy," which pertains to character and belief in oneself. (Walker and Avant, 1994, p.20).

Teamwork is conveyed as being vital for systems to work interdependently based on faith and acceptance of the members' character and competence. Working together requires one to know the basis of the other members' beliefs so that they can trust them to make sound and competent decisions.

The last antecedent of interdependence not- ed in concept analysis is **circumstance**, or **an event** or **situation**. A setting is required as a basis for the need of collaborative care. Individual entities must have a common goal to work closely together.

Consequences, on the other hand, are de- scribed by Walker and Avant (1994) as "events or incidents that occur as a result of the occurrence of the concept - in other words, the outcomes of the concept" (p. 167). Several consequences of interdependence are identified in this concept analysis. These were acknowledged as **less risk**, **knowledge**, and

transformation.

INTERDEPENDENCE HAS been known to create **transformation**, or **change**, which comes from a shared vision of a future. Multiple parts of a care team, for example, come together to make the whole and create transformation in the healthcare setting and change for the patient. It has been said that "positive relation- ships between nurses and physicians [are] ones in which nurses are able consistently to exercise professional judgement with positive results on the quality and outcomes of patient care" (Walker and Avant, p.28). A team should develop an equilibrium when working together.

Interdependence can be defined as the relation- ship between two or more parties that depend on each other for survival. Every part needs to con- tribute something to the other party to survive. It is not survival of the fittest.

In interdependence, there are more than one party involved. One-party will have resources that the other party needs to fulfil its requirement and the transfer works both ways. Both parties need to meet their need, and both can assist each other in fulfilling the need.

The relationship of dependence can be found almost everywhere because we need other people to survive. The relationship can be found with:

People – People depend on other people to

acquire the relevant resources to survive, this could be as basic as a farmer selling his potatoes to a client.

Regions – Different regions in a country depend on each other for resources. For example, one area can have fields that can be planted and harvest for food, and the other region would buy the food from the first region.

Nations – Different nations depend on one another for resources that are not available in their own nation. China could manufacture computer parts that are needed in America to build a computer.

Businesses – Businesses need each other to make money and to get resources. A company could depend on another business to supply them with the raw materials to produce their product.

These are just some of the interdependent relationships that can be found. Interdependence can be found all over society.

The nature of the relationship will determine the degree of interdependence. If the dependence is high and the one-party disappear, the other party will likely disappear as well. Client and supplier relationships is a form of interdependence. A client is dependent on the supplier to supply raw materials, and the supplier is dependent on the client to pay for these raw materials. The client will use the raw material to manufacture their product. The supplier will make use of the money to buy more raw materials.

Business can have interdependence relation- ship with financial institutions, where a business needs to loan money from the financial institutions to leverage itself to grow the business, and where the financial institution needs to lend money to be financially profitable.

Economic interdependence occurs when a party specializes in the fulfilment of a good or service and the parties need to trade with each other to meet the other's requirements.

Interdependence is also very crucial to the productivity of any workplace. It demystifies and diffuses **competition** in the workplace as all are focused on the desirable and acceptable final product which will bear the signature of everyone. Interdependence pronounces and showcases complementary spirit which announces team effort and encourages greater spirit of collaboration and participation. Inter- dependence is colorless in terms of race, gen- der, race, class, etc. because intellectual contribution to a product and the design thereof is colorless. All biases are shelved because the focus is on the final product, and if it will bear your signature, you want to have the best of it.

An allegory of a bouquet of flowers is a critical example to this: a bouquet of flowers is considered beautiful because of the many flowers and the multiple colors. Together, all the colors shine as exposed. The emphasis of the unique color is pronounced. The fewer the bouquet be- comes, the less the unique colors begin to be pronounced, the less visible, less noticeable, and less recognized the unique colors become, and eventually the less respected those colors become. The same is applicable to human beings. Our individual uniqueness is advertised when we are together than when we are apart. We are less likely noticeable when we are alone than when we are in the midst of others, especially, different colors. Remember that the flowers that formed the beautiful bouquet all agree to *relate*. Note that the relatedness agreement is in the fact that they are all flowers except that their colors are different. The same is applicable to humans when they agree to re- late in one spot. The different colors are more pronounced and beautiful together than when apart, and one by one. It is always more beautiful to be, and live, together than separate.

Chapter 4 - INTERACTION BETWEEN DEPENDENCE, INDEPENDENCE, AND INTERDEPENDENCE

Independence and Freedom

In some ways, independence is like freedom. One seeks independence from other people or things because one feels entrapped. Seeking independence, therefore, is like seeking freedom from what seems to be a bondage to someone. But we have heard several times that freedom is not quite free, so how can independence be freedom from bondage? Freedom comes with a price and sacrifice. The price one pays for freedom is the privilege to become self-sufficient, and in the real sense of it, humans are not self-sufficient because we are no designed that way. There is no one who can provide everything that he/she needs for himself or herself. There is reliance on another person for something, even if you must pay for it. Take a country that becomes or is proclaimed to be independent with huge celebration. Is there any time that that country is completely free? There is hardly any time. The country continues to rely on the support of the country from which she sought independence from. Apart from that, there is no true freedom or independence from the rest of the world. If there is trade and exchanges because there is no way to measure the economic sustainability of such a country. If a whole country cannot be completely free, there is the need to raise the question of the independence that is being sought by the humans, especially, the young people. If a country is completely left alone to fend for herself after seeking independence, it is equivalent to homicide because such a country would be subject to chaos, internal crises, political instability, economic instability, and the return to conditions worse than before the colonization or captivity, in the first place. It is for this reason that there are bilateral relations with other nations and international trade. One reason for such relations is the fact that no one country has all that it takes to be sustainable, hence there are exchanges between countries to ascertain their sustainability. This reliance on one another for sustainability is why inter- dependence is what should be the norm. With that, the continued interdependent existence is the right way to exist.

The interdependence relation complements where one is weak to where another strong. As earlier mentioned, no one country is all sufficient, hence where one country is stronger than the other, the other gets the necessary support for the weaker areas. It is riskier to rely on one country than to rely on several countries and on one another. When a country relates to only one other country, there is the risk of losing support from others, but it is not possible for that one country to have it all. It is better to rely on several because, more likely than not, a conglomeration of countries will be more dependable in terms of acquiring all that one will need. It is not possible for one country to meet the socio-economic and infrastructural needs of another country, but it is possible with several countries, by gleaning from here and there.

Independence and Younger Generation

THE YOUNGER GENERATION is so consumed by the concept of independence that they forget about maturity and preparedness for a life of independence. One who is not matured enough to sustain herself cannot be independent because independence means freedom from everything. You can imagine leaning on a substance that is not sturdy and steady. That is what it feels like to depend on oneself, especially, when one is not matured enough to sustain herself. A teenager is not matured enough to remain sustainable. The feeling of the teenager is that he/she is matured because that is what the society makes him/her feel. It only *seems* like, but it is not real. To be independent is to be independent in all things, not in some things; hence the significance of interdependence rather than independence. Independence is not truly independent, just like freedom is not freedom.

It is in submission that we can resist the tricks and the deception of independence. Submission here, for the teenager, is the willingness to submit to the instructions and guidance pro- vided by the parents. When the teenager exercises discipline to truly go through the up- bringing without rushing out into the world of independence, that is when they will be better equipped and matured to face the challenges of the world without getting burned. When one is broken and bruised emotionally, socially, and treated harshly for rushing out into the world of deception, it becomes more difficult to return to what one ran away from, and in the process of attempting to stake it out there, many run into different unwarranted situations that could have been prevented by endurance to be fully equipped or somewhat better equipped. Many of the teenagers pursuing independence at all costs feel killed emotionally, socially, financially, and thereby begin to engage in things that take their lives. Emotionally, because they are no longer stable to make wise decision; socially because they will feel left out and not able to fit into the circle of friends that they have, or even feel as successful as their friends. Financially, because they are not prepared to manage their finances to ascertain sustainability because they just followed peer pressure. Some eventually get destroyed. That element of pride that drives the desire to become independent will take a better part of them, and eventually, they get lost.

Independence and Peace

INDEPENDENCE GIVES rise to the concept of war because everyone living in isolation is so self-conscious of himself/herself to the extent that one begins to feel what the other person is not thinking. The fear of being attacked by the other person springs up such that one begins to self-protect against the other person. Living on edge in such an environment leads one to provoke the other because the thought of being attacked leads one to become more aggressive in their outlook and demeanor. With *independence* there is no concern about how destructive one can be with another person's belonging, but with *interdependence*, there would be thoughts of the impact of such destruction on their own well-being, especially, when one de- pends on the contribution of such a person. The destruction of that person's property will adversely affect your ability to function, hence the second thought before embarking on any destructive tendencies. Interdependence makes us think of collaboration before condemnation. The question to ask in interdependence relationship is: What impact will the destruction of that property have on my livelihood, but with independence, such question will not emanate because the effort is to fortify oneself against an attack that are, many times, unimaginable and fantasy and imaginary. In the world of interdependence, there is more collaboration and community consciousness than independence. In interdependence, the tendency is to seek the global good, not just the individual good. In interdependence, the focus is on self-improvement and communal improvement, because the good of the other is the good of the whole because everyone rubs off on one another for living. Take the issues with countries, for in- stance, several countries are living or existing in fear of the other because there are secret intelligence trying to figure out what the other is doing that might affect the existence of the other country. That is why we worry about espionage (spying) on one another. Every country is living independently due to the fear of the other. Can you imagine a world of interdependence, and what it would do to the world? There would be no fear of war, and there would be peace because one is *independent* and at the same time *dependent* on the other, which is summed as *interdependence*. There would be no need for spying on one another because the good of one is the good of the other. Come to think of it, several countries display the concept of interdependence, even though, they will not agree to it, by forming alliances. The concept of forming alliance is the concept that we are all independent but not self-sufficient, hence one country would need the other to be an eye for the other. Nations that are allies can visit one another, trade, collaborate, etc., based on diplomatic relations. That is the power of interdependence. It does not rob any country of their supposed independence but affirms that no one country is self-sufficient. One country has what the other needs in addition to what the country already has, and vice versa. The deceit of independence makes nations to spend more money on preparing to defend themselves than on the livelihood of the indigenes and citizens. In- stead of shining light on independence, we can promote interdependence. As a matter of fact, the more we shine light on interdependence, the dimmer the light of independence will be- come. That means that the more we will realize that we are a combination of dependence and independence. People have said that life is contribution, and no concept affirms that more than interdependence. Independence does not promote the idea of contribution because the focus is on self. Dependence does not promote that idea either because it is focused on the other person only. Interdependence promotes and supports that idea of contribution because you are contributing to a cause that affects all, and all will benefit from it. If I am dependent, I do not have anything to contribute. If I am independent, I could care less about other people. If I am interdependent, I have that obligation to make something happen for the collective good, and so do others have the obligation to- wards me.

Interdependence does better to the society be- cause it will reduce or eliminate crime. This is so because it would be difficult to commit murder knowing that I depend on that person as much as he/she depends on me. Killing him or her would put a heavy dent on my livelihood, so there would be the deterrent to killing the person. There would be no homicide because doing so to oneself will leave a gaping hole in the community and society. There would be no need for robbery and kidnapping because I would be robbing somebody who is an extension of me; or kidnapping a person that makes me complete as a human being.

Independence Day Celebration

AS I REFLECT ON WHAT we refer to as 'Independence Day' celebration of a country, I have come to discover that there is really nothing like independence. When a country is declared independent what is being done is simply that the country is ready to switch from being completely dependent on another to being able to becoming sustainable to a certain extent. It is almost like a paradigm shift from dependence to independence, but in the real sense it is a declaration that the country is ready to begin to make contribution to human existence, which is a display of some level of maturity to fend for herself to a certain extent and become a contributor as well. Instead of absolute dependence, the country can now bear some of the burden.

Have you noted that it is when a country be- comes independent that she can be welcomed into the United Nations; she can be allowed to participate in the world or international trade; participate in the Olympics and other world sporting events? That is why it is graduation from total dependence to the level of interdependence. It is graduation from the level of help to assistance. Help connotes helpless- ness because you do not have anything. Assistance connotes some level of ability to do some of the required things but only needing some support to complete it. When a country is declared independent it is truly a declaration that she can provide some things for herself now, and that she can have some leaders who can lead, while at the same time would need limited guidance and direction.

Look at the case of Allied Nations, it is the drive for interdependence that led to that creation. How about NATO? The idea of forming a coalition is the notion that those countries can- not operate independently, hence they form a league of interdependent nations. The idea is that each country by herself cannot have the needed security, hence an ally is needed to assist or even be an eye to watch for any noise that would threaten their peace. Coalition is interdependence for fortification. Any way one looks at it, interdependence is what the countries practice rather than independence. Declaration of independence, therefore, is acceptance into the category of alliance of inter-dependent nations. It is the acceptance into the world economy platform where she can participate in trade. It is an acceptance into the con- glomeration of nations that want to strengthen themselves. The country that is left out and seeking to be independent always appears to be a common enemy. Interdependence is valued over independence. In terms of trade, all nations desire to have some form of economic relationship with the others because the idea of self-sufficiency is considered foreign since no one country can fend for herself. Per the business standpoint, it was stated that interdependence meant the dependence of entities on one another such as people or countries. For example, over the years, the United States' interdependence on other countries for imports has grown to record levels.

I will propose that we should begin to address Independence Day as Interdependence Day because that is what it really is. Interdependence brings out the best in the nations. Even when there is competition, it is competition in the pursuit of excellence for better contribution to the welfare of the whole world.

Interdependence and Marriage and Family

AS A PART OF GROWTH of the human anatomy, every single person who is of the marriageable age desires to be married because there are certain needs of humanity that can no longer be met by the individual even if they have been with their families. They desire to be independent first, but they still want that interdependent relationship because of the fulfilment of the true humanity. They move from the level of dependence on parents and family, to independence, but they are not satisfied until they reach that point of interdependence which brings out the true meaning of humanity. This fulfilment, when lacking in many peoples' lives leads to depression, loneliness, isolation, etc. Come to think of it, the main driving factor of marriage is interdependence. The purpose of marriage is, there- fore, that fulfilment which can only be defined by interdependence.

If the purpose of marriage is other than interde- pendence, then it is not marriage. The reason for going into marriage is because there are things that you could not provide for yourself but that you need. Whether you are the male (husband) or female (wife), those things that you need to grant you the fulfilment in life and be- come complete are in the other person, hence you get into a relationship. You will notice that whenever you do anything short of it, you begin to have a lot of conflicts and frictions in your family. It is not marriage when you decide to be independent or be dependent. It must be interdependence. Notice that the description is that marriage is about bringing the full realization and fulfilment of who you are, then how can you pursue independence and eschew interdependence? No marriage can survive as long as it should when there is emphasis on anything short of interdependence. In a marriage where there is only one person providing for the family, dependence, there will arise frictions and conflicts as the provider (husband or wife) will soon begin to have ideas as the only one making any meaningful contribution to the well-being of the family. If not personally generated ideas, other external influences will soon pressurize the person into such ideas. Even when it is the agreement of the couples to have one person to provide for all, the person that is not the pro- vider will soon begin to feel unfulfilled because it is the human fundamental right that a human being should work, and failure to do that will result in other adverse implications emanating from the unfulfillment in living. The reason is because no one wants to live a life of dependence all their life, which is where the desire for independence comes from. In the case of in- dependence where either of the couple works but do not make contribution to the well-being of the family together, but on their own, there will also be conflict as there will arise the notion of secrecy which should not exist in an interdependent relationship. The idea of secrecy in a relationship will lead to the feelings of in- equality and unfairness as there will be feelings of contributing more than necessary by either person. Nothing gives rise to quarrels in a family more quickly than financial conflicts, hence it is better to dwell on the axis of interdependence where there are no secrecies but transparency. A life of independence gives rise to a life of secrets. A life of secrets leads to a life of fear. A life of fear leads to a life of insecurity. A life of insecurity leads to a life of amassing defense mechanisms. A life of defense mechanisms leads to a life of self-defense. The resulting end is self-fortification against one's spouse and family members whom one ought to live happily with. If one spouse fortifies himself/her- self against the other, there is a breakdown in marriage, and marriage seizes to be marriage but a war zone. What even makes it worse is the even where there is no such fortification, the assumption of the collection of weapons of mass destruction that do not exist can even create so much conflict and destruction of the marriage. Secrets lead to unwarranted tensions due to the uncertainties, and therefore, the thought of elimination to avoid being the victim. The fact is that such tension could be based on fear that is non-existent. That is why marriage must dwell on interdependence, not independence or dependence. Where there is openness, there is transparency. Where there is transparency, there is exposure to the needs of each other which enhances complementing values to support each other. Self-exposure settles conflicts, assumptions, insinuations, doubts, etc. Self-exposure enhances interdependence as it exposes self-insufficiency. The same goes for a family that is interdependent where no secrets exist. Show me a family that is interdependent in her values, and I will show you a family that lives happily ever, or better yet, a family that is joyful. It is the fear of secrecy that gives rise to conflict because each one is uncertain about the other and his/her plans. The secrecies are too much and many that no one wants to relax, but always living on the edge and in tension for the uncertainties that might unexpectedly come.

Taking the issue of independence, a notch further, this concept in marriage leads to oppression and intimidation. The fact that one feels that the secrecy of the other person might result in who strikes first leads one to ensure that the power is reduced or destroyed by ensuring that the person is weakened to the point of not being able to carry out or execute their plans, whether it is imaginary or not. This leads to intimidation, and sometimes, oppression of the other partner. Where there is interdependence, hardly will such exist because there is collabo- ration, cooperation, and unity of purpose.

There is no marriage that can survive in the atmosphere of independence and dependence. Independence is deception in marriage. Dependence is burdensome in marriage, but I make bold to state that any marriage can survive, as a matter of fact, can thrive in the atmosphere of interdependence because that is what marriage is by description. It is relying on each other, not on oneself and not on the other. Where such reliance exists, it can only be possible under the auspices of communication. In the atmosphere of interdependence, no one can survive without communication. You either communicate to make your needs known or you communicate to meet the needs of the other.

In marriage, the concept of independence leads one to withdrawal. Withdrawal syndrome is very dangerous in any relationship, especially, marriage. As a matter of fact, withdrawal is a killer of relationship as there will be no more relationship. It takes two to be in a relationship so when one withdraws, relationship ceases. The concept of interdependence makes marriage relationships to thrive. Very close to marriage is family. The concept of family is togetherness, but it is the concept of interdependence instead of independence.

Can marriage become a utopian venture where one does not have the ability to think on their own? That is not what is being discussed here, but that where there is interdependence, such thinking and ideas are shared, discussed, de- liberated upon, and executed cooperatively.

Interdependence and The Composition of Human Being

THE TRIPARTITE NATURE of a human being sup- ports interdependence existence: spirit, soul, and body. That is the order of existence. These three elements cannot exist independent of the other and cannot be completely dependent on another but exist interdependently. No human being can exist by the spirit only; soul only; and body only, hence all three are active participants in what makes one to be a human being. Each of them is a dimension of the human body.

Interdependence and the Human Existence

IT IS ALMOST INCONCEIVABLE to think of human existence without reflecting on the concept of interdependence and the contributory values. We even have many adages that cause one to reflect on this, such as: *"two heads are bet- ter than one"*. Two heads are better than one means that when two people work together they are more likely to solve a problem than one person doing it alone. "Two heads are better than one, not because either is infallible, but because they are unlikely to go wrong in the same direction." — C.S. Lewis

It is not just any two heads but two working, independent, dependent and productive heads are better than one. This is the idea that those two separate heads are functional such that when they are together, the fusion creates a dynamic outflow of resources, and such fusion is possible only in the atmosphere of interdependence which allows for the sharing of knowledge, insights, skills, wisdom that leads to a synergetic display of intelligence. In ad- dressing business strategy, Richard Branson states that, "Two heads are definitely better than one and by brainstorming as a team and sourcing ideas from each other, you have a better chance of coming up with a strategy that will allow your business to overcome a setback or challenge."

One mind can think only of its own questions; it rarely surprises itself. - *Orson Scott Card (1996). It takes a seriously minded person to question his/her own mind. It takes some serious discipline to engage questioning oneself, but it is easier to question oneself in the atmosphere of other people. There is a need to check one's thinking for evaluation and analysis to determine the focus. The significance of interdependence lies in the fact that,* many *heads are inevitably better than one.*

The need for interdependence of human beings can also be seen in this African proverb: *"If you want to go fast, go alone. If you want to go far, go together."* By going far, it is not just in distance but also far in accomplishments and achievements. While going fast is a great thing, it is not always the best especially when you are going alone. It is possible that many things could be left behind and be unnoticed, and such things could cause danger and thereby hamper one's steady progress. On the other hand, when you go with others the journey is more pleasurable, and more importantly, the checks and balances can be the reason for a steady progress and successful adventure. Those with you will ensure that no stone is left unturned, as such unturned stones could have an impact on them positively or negatively.

Concurrently, the African proverb: "*Go the way that many people go; if you go alone, you will have reason to lament*", illustrates that an individual's cultural security lies in the personal association and contributive role played within a group.

These sayings and many others clearly state that humans are created or born for interdependence if they are to achieve much and be fulfilled. If anyone wants to go far in accomplishments, achievements, fulfillments and successes in life, the main way is interdependence. The person will have to take the "*Interdependence Route or Boulevard*". Going fast might not lay the foundation that will fortify the desired outcome, and hence it will be temporarily lived. When it involves another person or other people with whom one collaborates, it is certain that other perspectives will come to bear, and therefore the dynamics will change which will lead to greater success. As a matter of fact, the perspectives of all people will lay a very formidable foundation. Working together opens the doors of possibilities that will raise questions which one person might not see. There will be no surprises at all if you go alone, but there will be an opportunity to make it better as any potential errors might be exposed, and besides, the other views will shine brighter light for greater success.

For a nation that wants to go far in development,

there is no doubt that collaboration by means of interdependence is the way to go rather than going independently. No nation that is independent of others will ever go far enough because there would be no comparison, no collaboration, and no healthy competition. What- ever such a nation produces, or manufactures would be the only thing known to such a nation. If it is the poorest quality, that is what the nation is stuck with. Where interdependence exists, the nations will be able to share ideological differences that will enhance the products being manufactured and ensure quality control and international standard. The fact also remains that even if a nation is a producer of a product, without the patronage of other nations, the product will not go far as to attract international recognition.

Interdependence in human existence leads to wisdom living and avoidance of life's con- sequences. When people are interdependent, there are opportunities to learn from the errors and corrections of others and by so doing enable others to desist from such costly mistakes. A life of independence would want to experience all sorts of errors and suffer consequences which might have easily been avoided. Why choose to live by consequences instead of by wisdom? Some of the consequences could be costly and deadly. Parents undoubtedly lament the refusal of their young ones to heed the wisdom of their instructions, knowing that those instructions would spare them unnecessary emotional and physical pain.

When it comes to interdependence of one age on the other, the younger generation must cultivate the attitude of learning from the wisdom of the older ones, while the older ones learn about new technological systems from the younger ones. The interdependent ability of one age and the other can only lead to greater achievements in existence as no one will be left behind and no finger will get burnt. It can only lead to growth in wisdom for all as no one would be useless and no one will be indispensable. The young and the old must interdepend to live life to the fullest.

Independence and the Marketplace

THE CONCEPT OF INDEPENDENCE seems to do everything but what we attempt to promote in the workplace. One of the major things that make a company to fail is the idea of working in silos. Companies spend a lot of money and commit a lot of resources to ensure that there is a break- down in working in silos. The reason is that we believe that the morale of the staff (the most valuable asset) is key to the productivity of such an entity. Hence the leadership promotes team efforts and communication. We promote outward mindset because we want to promote interdependence, but we also want to be independent at the same time which really works against the concept of collaboration. While it is not the intent of the establishment, it is the reality of independence versus interdependence. The concept of outward mindset encourages reaching out to other people and seems to pro- mote the idea of seeing things from the perspectives of other people as well. But this is what independence does not encourage because it is all about I, myself, and me. The craving for independence is more about doing things by myself. Establishments encourage communication which can lead to profitability

, but independence seems to be in the way of this vital virtue at work. In the Marketplace the concept of independence seems to encourage unhealthy competition among staff because everyone would be fixated on attempting to undo the other instead of working hard to get every- one on board. There will be hoarding of information from one another because no would be encouraged to share their knowledge for fear that it could be a weapon against them by their competitor. Instead of seeing the fellow staff as colleagues they are seen as competitors. The concept of colleague at work is the concept of interdependence which promotes the fact that if we all get recognized, then we all get the cred- it together. As a matter of fact, the concept of interdependence should be a key performance index in the marketplace. If there should be kudos for anyone in the marketplace, it should be for everyone on that team because no one individual can start a process and complete it without the contributions and efforts of others on that team or in that department. Come to think of it even other people from other departments must have played a role which amount- ed to the production of that which attracted the kudos. Take for instance: If the Payroll Unit does not process payroll, you will probably not be of the mind to work well for the product. If the Accounting department does not pay for the contractor of the supply for the product, the materials, or elements to produce the product would have been missing. If the leadership did not encourage the process, there probably would have been no incentive or impetus for the work which produced that product. Every person must share in the product which attract- ed the kudos. In the case of a manufacturing establishment, everyone from the driver to the security guard to the cleaner, and so on, should share in the kudos. This is how the concept of interdependence works. The interconnectedness that exists in the working dynamics of in- terdependence makes it impossible to lift one person without the others being lifted. The in- terconnectedness is like the fishing net, which is all connected, and you cannot lift one part without everything else along its way coming up with it. The recognition of one is the recognition of all, and consequently, it diminishes, and there is high probability that it will eliminate inequity.

Chapter 5 - The Philosophical/Social Discourses on Interdependence

Philosophically, more specifically, phenomeno- logically, the realization of oneself is based on the self-consciousness which emanates from the stages of life as one grows up. This aware- ness does not rob one of his/her uniqueness as a person, instead it affirms it. There are stages of growth from birth to self-consciousness. I know that I am a human being because I see other human beings like me. This promotes interdependence. The essence and determinant of our human existence is challenged when we promote independence at the expense of inter- dependence. Show me your humanity and I will show you interdependence.

Independence promotes dichotomy between human beings in different realities. Interdependence promotes bifurcation of the different dimensions of the same reality. Instead of a dichotomy between dependence and independence, there ought to be a continuum between the two which could be labelled as Interdependence. There should be an interpenetration of independence and dependence which could be rightly termed: *interdependence.*

The concept of dependence and independence

has created a dilemma in human existence because the approach has been that of realms. Realms seem to be viewed as two separate sectors that do not interact for any reason. Instead of the realm model, the approach of dimension should be considered and applied to the discourses. While realm-model bifurcates, dimension-model distinguishes. Dimension-model emphasizes the simultaneity of human existence and would support the idea of dependence and independence as simultaneous existence. Reality and experience of human existence emphasize that there is simultaneously interpenetration between the dependence and independence of human existence. Dimension-model will help to overcome the dualism of dependence and independence. The approach of dimension-model calls for a continuum instead of bifurcation or dichotomy. Comparatively, human is a continuum of dimensions from the physical to the moral to the personal and the spiritual. The concept ap- pears to be hierarchical in nature, but there is a dynamic continuum between them. The *physical* dimension is the aspect of experience in which one is aware of the material world. The *moral* dimension deals with one's awareness of other persons. *Personal* dimension deals with one's awareness of himself as a person. Self-awareness is what empirically separates human from other animals. Self-awareness is mediated by the physical and moral dimensions. The necessity of the physical dimension for self-awareness should be self-explanatory. Self-awareness because of the relatedness with others leads one to the *spiritual* dimension of the tacit level. Self-awareness or conscious- ness is what affirms that human is indeed human because of the interaction with others, therefore, the concept of striving to be independent from others, does not make sense. The idea of dimension-model makes sense as it emphasizes that one human being is an extension of the other, which further emphasizes that interdependence is what we should aspire for, not independence from the other. When one begins to feel an overwhelming sense of self sufficiency without others, one is proud.

Noumena and Phenomena

TO BORROW FROM THE language of the eighteenth-century Philosopher, Immanuel Kant, the concept of the noumena and phenomena is very relevant for our understanding of the dis- course of interdependence. Kant describes the 'noumena' as '*things in themselves*', and the 'phenomena' as '*things as they appear*.' It is not to simplify the concept as described by Kant, but I find relevance to the discussion. While for Immanuel Kant, the concept was used to stress the dichotomy that pervades our existence, I see the concept as deeper. There is a great difference between perceiving humans '*as they are in themselves*' and '*as they appear*'. When you see humans as they appear, you are deal- ing with the temporal, ephemeral, and one is subject to interpretations that do not add value to anything because the person doing the perceiving cannot claim any good knowledge of the person, but simply being judgmental. There is a glaring difference when one sees a human being as he/she is in himself/herself. This means that there are values and virtues that are inherent that are being studied and understood. To see one as he/she appears is just at face value. In the noumena frame of mind, you see values and virtues which supersede mere appearance. The force to see a human as he/she appears is what leads some humans to maltreat others, to lose respect for the other, but when you see that what you are seeing (outwardly) is just the covering, and that there are better things beneath the covering that can make a great difference in life, then the reaction will change. The thing that is beneath the covering is of eternal value, and such things require some level of intentionality to discover the person as in himself/herself. Many fail to take that extra effort to know the person as he/ she is in himself/herself; and it is such failure that makes us to dwell at that ephemeral state of the phenomena frame of mind. The phenomenal frame of mind is deceptive and does not bring out those eternal values that truly define and describe the essence of the human, but the noumena gives a better picture because one is seeing beyond the ephemeral, and more than the eyes can behold. The saying: 'Beauty is in the eye of the beholder' takes a different interpretation because 'beauty is in valuing the noumena (seeing the person in himself), and not in focusing on the adorning of the noumena.' I consider the noumena frame of mind as the bedrock of interdependence because it draws one to realize the unique thing in the person, and the eternal values that are embedded in such a person that could contribute so much to fix the defects in other humans. Every human needs to cultivate the noumena frame of mind because it promotes that unique quality of mutual respect in humanity.

Interdependence and Diversity

DIVERSITY IS A WORD that we do not really grasp the meaning because it is abstract to us. Diversity cannot be understood outside of what I call 'relatedness'. It is in the demonstration of what brings out diversity that its meaning resides. It is not in relationships but relatedness.

To be related, there need not be any emotion- al attachment to the like-mindedness which attract interdependence. There is the level of acknowledgment of the existence of the like- minded or look alike depending on what you are seeing or want to see. What do you see? Do you see all people being the same but in different kinds of ways as human beings?

Do you see other people as staff like you at the organization, or do you see supervisor and supervisee dichotomy thereby creating the class of superior and inferior? It is easier to relate with people you identify as being like you than not. This is not talking of relationship because it is not a promise and it has no emotional attachment, as they are not ingredients of relatedness. There is no need for any effort in relatedness as in relationship. It is unforced rhythm of our makeup as humans. We do not have to strife to relate because there are no logical rea- sons to do it. One thing to know is that when an effort is required, then it creates room for fraud, conditionality, compromise, manipulation, pain, deception, pretense, avoidance, adulteration, fake, counterfeit, and disappointment. These are the elements that cause relationships to fail, and they have no room in relatedness be- cause relatedness ought to come naturally. If you must struggle to relate, then what do you resemble: the same or another? The mere fact that you recognize the sameness in all human beings is the necessary element of related- ness. Why can we not relate? Unity in diversity is in the relatedness between and among us. Diversity is only pronounced when it comes together, just as colors and different shapes are more pronounced in togetherness. A color on its own is not as beautiful as when it is a part of a group of colors or a conglomeration of colors. A shape is not as pronounced by itself as when amid other shapes. In the same way, the unity of diversity is evident in togetherness, and togetherness is only pronounced in relatedness, not relationships.

The description of diversity is embedded in the

practicality of relatedness, and its sustenance is in the effortless display of such rhythmic existence. No extra effort required. No need to protect oneself from the other. It comes with intentionality and determination. In the world of diversity, everyone matters and counts because everyone can relate to the other person. If it is not so, then it is not diversity. Acknowledgement of diversity is not diversity, but the relatedness which is what identifies that essence.

At the core of interdependence is diversity. Inherent in interdependence is the concept of diversity. Without the recognition of diversity, there can be no interdependence. Diversity is not limited to race, but it is in everything, even including identical twins and siblings, because there is something different about them and in them. The diversity in a family is sealed by the bloodline but it does not make it any stronger than the general bond that should exist among human beings. There is diversity even among the same races.

Difference Between Relation and Relationship It is in human nature to seek connections with other people, animals or even things. This connection can be defined in terms of a relationship or relation. A relationship is link between two subjects or human beings and is used informally. In the case of human beings, a relationship is a special connection that is either bound by blood or marriage or by a unique union. A relation is formally used to identify the link between two groups, countries, or elements.

What is Relation?

THE TERM RELATION IS used formally to identify different connections. Relation can also be used to identify similarities between two different things. Relation is used in a formal context to define the connection between groups of people, similarity between different elements and to establish a personal account of different subjects. Relation can be used to describe similarities and the sameness of two different people, things, or countries. Relationship cannot be used in the same way. Relationship cannot be used as a reference point. Relation is used in a formal context,

What is a Relationship?

A RELATIONSHIP IS A term used informally to identify linkages or various connections between people or item and describes how they are associated. Relationship is used informally to describe connections and linkages that exist between people or elements, to describe behavior and classify a couple's intention.

Summary of Relation vs. a Relationship

- Both the terms, relation and relationship, help to describe some form of connection or link between groups of people, two individuals, non-living things, or elements.

- Relation is used informally while the term relationship is used in a formal context.

- Relation can be used to identify the similarities between two different things, groups of people or individuals.

- Relation is also used as a reference point.

- Relationship describes the connection between people, their behavior towards each other and a couple's mutual status.

IN A RELATIONSHIP, there is an emotional attachment, if not it is not a relationship. In a relation, or what I call relatedness, there is no need for any emotional expression or attachment as it is a recognition of the similarities that exist between people. It is just a recognition of the likes among a group of people. There can be the relatedness of a group of people, but relationship for a group of people is unrealistic as there are many things that cannot be fully expressed. Elements such as trust will be far- fetched. Trust is not needed in relatedness or relation, but no relationship is complete without trust. There is no need for any expectation of reliance or reliability in relation or relatedness, but it can strengthen that relatedness if present, however, reliability is a great ingredient to establish and sustain a relationship. Hence, interdependence is grounded in relatedness, not relationship because it is not possible to be in a relationship with all human beings, but interdependence must exist in the celebration of our humanity and humaneness. You can imagine how much and how far this can go in the peaceful co-existence that we so desire.

Uniformity and Unity

THE IDEA OF UNITY CONNOTES uniformity which further emphasizes the fact that there is need for interdependence. It must be stated, however, that unity is not uniformity, that is, we are not all supposed to be one and the same to be united. As a matter of fact, if we are uniform, there is no need to promote unity because we would be already united. It is a colorless society. It is a title-less society. Interdependence thrives in a colorless and title-less society because everything is seen from the lens of humanity rather than from the lens of categorization/classism. From the perspective of humanity, we will not categorize people by tribe, region, section, tribe or even race. Titles like Latina, Asian, Native American, African American will be heavily de-emphasized. Though there are heavy cultural values attached to those classification, those cultural values will not be what will give the people identity. It is the attachment of those titles to their identity that creates the systemic racism which further classifies them as inferior. When we value everyone's contribution, we will not diminish anyone as inferior.

Anyone who tries to survive alone will suffer because it is anti-design of humanity to do that. Humans are packaged to interact with one an- other. As a matter of fact, anything outside of that is a denial of one's actualization of their being. Your survival level is measured by the level of interaction with other people. It is what makes the difference between survival and ex- tinction in any situation that one finds himself or herself. It is in the place of independence where one finds such phraseology as: "*Survival of the Fittest*", but in the place of interdependence is where you find or discover that the fittest is the one who interacts with other people not the one who goes it alone. The journey of life is not an alone journey.

The realization or actualization of your independence is to the extent of your participation in the interdependence circle. You will realize that you are independent, but you can participate in the interdependent world as a viable and crucial contributor to the well-being of humanity. In the realm of interdependence, you can also realize that you are dependent when you do not make any contribution at all because you will feel that you are a parasite, feeding on other people's terms. Where interdependence is missing, independence and dependence can- not stand on their own for long.

Vulnerability As the Doorway to Connectivity

AS DISCUSSED UNDER the characteristics of 'dependence', vulnerability is one of its negative aspects, but vulnerability can also be the door- way to connectivity. Every human being is a vulnerable being because he/she is exposed to some form of danger, and the only thing that can bail such a person out is assistance from another person. That is what gives rise to the connection in vulnerability. The thing that connects us is the vulnerable nature that we have. The vulnerability is not just profound in the visible or physical but could be manifested in the emotional and social state. The concept of connection is feasible in the sense of the need to fill that gap in human nature that creates the vulnerability. Because this is a general phenomenon, even if one denies, it only begs for interdependency of human beings. Interdependence is connection; hence vulnerability is the doorway to interdependence. Take for instance a frail older person needing assistance by connecting with a younger person, preferably, a little child. The older human being is vulnerable because of his frail condition and would need the support of the young person. At the same time, the young person would also be vulnerable in the world ahead of him and would need guidance and counsel from the older man on how to navigate the future. If there is no vulnerability, there will be no need for connection. The realization of the vulnerable state of human forces the concept of connection and interdependence. If one remains and practices independence, the vulnerable state will lead to other debilitating conditions. If one chooses dependence, then other parties will feel over- whelmed and get frustrated and discouraged, but if each person has something to gain from the connection, it will be sweeter to engage. There is a symbiotic relation that is established to deal with the vulnerable state of humans.

Beautiful versus Looks Good

WHEN ONE WEARS A SINGLE-colored shirt or dress, one hears the comment: *That looks good on you*; but when it is multi-colored, one hears comments like: *That is beautiful.* Diversity that embraces relatedness is beautiful to behold. The secret is in relatedness of the colors which are not in a relationship. The challenge we have in life is that we think that we must be in a relationship before we can flow together. It is a myth because those are two different things. If one beholds a color, it should intrigue such a person to discover more about that color or the difference that one notices. The one thing that inhibits the pursuit of knowledge is the idea that nobody is like me, and that I am better than the other persons. Such thoughts shut down one from the quest for knowledge because one would not even want or have the incentive to find out more about such a person. In the world of interdependence, beauty is pronounced.

The Uniqueness of a Person

EVERYONE HAS SOME ELEMENT of uniqueness in him/her, but the fact is that, on its own, unique- ness does not really mean anything unless it is in relation to the general. It is in the multitude of uniqueness that each individual uniqueness is revealed and pronounced and relevant, just like an individual on their own has no relevance unless it is in relation to other individuals. The relation to the multitude is in the acceptance of the like beings as having something in com- mon. Using the allegory of the bouquet of flowers, each flower has some level of uniqueness, but it is in the relation of one to the other that the beauty of each is pronounced and emphasized. We can take our cue from the bouquet of flowers, that, for us to function and perform well, as human beings, we need each other and one another to bring the best out of each of us, and that can only be possible in the realm of interdependence. Independence will deny us that effect which consequently will isolate us. Dependence will not do it because it will make us to not showcase our own potential. Interdependence will do it because we will have the opportunity to identify our own uniqueness and contribute to the good of the populace. It is worth noting that no uniqueness of an individual outshines the uniqueness of others because the uniqueness of one person complements that of other people, and together the output is beautiful. That means that, on your own, your uniqueness does not get a mention unless it is in reference and relation to the conglomeration of uniqueness. Even in the workplace or marketplace, the product of a department or unit is because of the conglomeration of the different talents, duties, and responsibilities of each member of the department. The failure of one person to perform always leads to a failure, a delay or even a lack of product. Each member of the department must perform his/her duty and usually, the output is beautiful, but only in the face of acceptance of the contributions from the different members who were hired to fulfill that role. The same way we interdepend on one another, in the marketplace, to produce beautiful reports, so also are we expected to interdepend, in our existence, to make the world a better place. Your uniqueness, kept to your- self in the name of independence (isolation), does not promote the good of anyone and of our existence. Your uniqueness, suppressed in the name of dependence, (the lack of sharing and contributing), leads to overburdening of the person one is dependent on, and produces no good thing to the existence of humanity.

A Beautiful Story of Interdependence - Two Friends with Defects

THIS STORY IS TOLD by **Helen Murphy**, published on July 12, 2019.

Woman Who Can't Walk Teams Up with Blind Man for Hiking Adventures: 'He's the Legs, I'm the Eyes'

"If I fall, she falls," says Trevor Hahn, who is working with his pal Melanie Knecht to climb a fourteener — a mountain that's over 14,000 feet

Melanie Knecht and Trevor Hahn are showing the world the meaning of teamwork.

The hiking buddies, who both live in Fort Collins, Col- orado, have worked together to navigate Colorado mountains and trails ever since meeting at an adaptive exercise class, according to Good Morning Amer-[1] ica[2]. Knecht was born with Spina bifida[3] and uses a wheelchair to get around, and Hahn lost his eyesight due to glaucoma five years ago, the outlet reports.

After they bonded over their love of the outdoors, the pair started hiking together. During their treks, Hahn carries Knecht in a secure harness on his back, and Knecht gives verbal directions to guide him along the trail.

---

1. *https://www.goodmorningamerica.com/living/story/walk-lost-sight-reaching-heights-64192036*
2. *https://www.goodmorningamerica.com/living/story/walk-lost-sight-reaching-heights-64192036*
3. *https://people.com/human-interest/2-year-old-boy-spina-bifida-crawls-dads-invention-frog/*

"It just seemed like common sense," Knecht told GMA[4] in a recent interview. "He's the legs, I'm the eyes — boom! Together, we're the dream team."

"This way, we both have purpose and this huge re- sponsibility," Hahn added to 5280[5].

The pair documents their adventures on the Instagram account @hiking_with_sight[6]. On their profile, Knecht and Hahn describe their hikes as "a journey of purpose between two friends, one who cannot see and one who cannot walk."

Hahn told KDVR[7] that he continued to hike and climb mountains after he lost his eyesight, relying on adaptive techniques like following the sound of a bell.

"But it didn't really give me a purpose. Like, I was just following this bell," he said. "It would be really cool if I could have a purpose on the trail."

After hiking with Knecht, Hahn now feels like he has his purpose.

"It made me so happy to help someone experience what I've been able to experience my whole life," he told GMA. "Just getting on top of a mountain, a car can't get to it, you just feel that sense of accomplishment. The best part is being able to make her smile. That gives me purpose."

For Knecht, the best part of their hikes is being able to leave her wheelchair behind.

"I've been in a wheelchair my whole life, and it's an amazing feeling to leave it literally miles behind on the trail," Knecht told the outlet. "I even couldn't get in it if I wanted to, and that's a great feeling."

"There's definitely a learning curve," Knecht told 5280 of training for their biggest climb yet, adding that the most important skill for the pair is "communication."

"I'm trying to warn him about obstacles far in advance, but also tell him if he's about to trip on a rock or root in that moment," she said. "I have to interrupt myself to give directions."

"It takes a lot of teamwork," Hahn added. "If I fall, she falls."

Knecht also shared her advice for others who have friends with disabilities.

---

4. *https://www.goodmorningamerica.com/living/story/walk-lost-sight-reaching-heights-64192036*

5. *https://www.5280.com/2019/06/hiking-with-sight-duo-plans-to-conquer-a-fourteener/*

6. *https://www.instagram.com/hiking_with_sight/?utm_source=ig_embed*

7. *https://kdvr.com/2019/06/04/colorado-hikers-one-blind-and-one-in-a-wheelchair-use-their-strengths-to-help-each-other-climb-mountains/*

"Ask questions of people with disabilities, to see what they like and what they want to do," she told GMA. "Don't not include them because you think they won't be able to do something."

You can imagine how two people with such defects will hike miles of mountain trails. The duo realized they were one another's perfect complement for enjoying the wilderness: The male, who was blind was strong enough to carry the female who had the debilitating defect but could see. The female member of the duo would give direction to the male who could not see. They were a 'dream team.'

The members of this dream team did not engage in pity party, seeking empathy for their conditions. Their focus was on being human by recognizing that they each had something to contribute to human existence. There was no rancor about what the other person lacked or not. They did not engage in being prideful because of comparison to the other. Their focus was on cheerfully giving of their gifts in service to the other persons and recognizing how they could each better the life of the other by their collaboration. In this collaboration, the male counterpart did not have so much to gain because he could not enjoy the trail and the scenery of the trail, but he relied on the explanation of the female counterpart and enjoyed the imagination. He was not concerned about the burden of carrying the female counterpart on his back but was just focused on the fact that he could affect the life of the other and the other could affect his own life.

In a life of interdependence, there is purpose. A life on independence does not really have any purpose or fulfilling purpose. A life of dependence has no purpose other than the purpose of the person on whom you depend. Just like Trevor Hahn says it here, using adaptive techniques might help but it does not give purpose and meaning to the adventure. So, it is with life's journey that is embarked upon alone or one that is dependent on another person.

Like Knecht said, there is nothing better than relating to and with a human being, even in the face of the technological advancement that we tend to rely on and make our sustenance. The best part of the hiking was leaving the wheel- chair at home and connecting with another being like her.

There is a general or common love for all hu- man beings, and that is the love of living and enjoying life. The realization and fulfilment of this common love comes with a sense of dependence and security and reliance, as was the case with the dream team. The full realization of their love could not come to fruition with adaptive techniques, but with a fellow human. Adaptive techniques could not explain what was being seen neither could it prevent from a fall or explain the beauty of the scenery, but the enjoyment of the trail was by the explanation. The common love for living that humans want to enjoy can be found in the interaction between and among human beings. This interaction can be more evidently experienced in inter- dependence. While other adaptive techniques might be designed to assist one, there can be no purpose and fulfilment. What is the purpose of going on a trail if there is no purpose? So also, one can say, what is life and living if there is no purpose? The trail can be compared to the journey of life. To have this sense of purpose and enjoyment, it takes responsibility and reliance. Just like with the dream team, in the journey through life, no one person is complete by themselves, but you could be somebody's eyes, and another person could be your legs. The question is: can you rely upon the person who is your eyes to give you great direction? Can you be relied upon to give security to that person whose legs you are?

As was expressed by the dream team, communication is key in the interdependence world. We need to know that in this communication are reliance, responsibility, security, prevention, guidance, and so on. It is not just ordinary communication by speaking because when wrongly given or with lack of clarity, disaster will strike. Hahn says that if I fall, she falls, and that depends on the clarity with which she communicates. Communication in the world of interdependence must be clear to avoid confusion and being confounded.

I must draw our attention to the statement by Knecht: "I have to interrupt myself to give directions." In the world of interdependence, I, me, myself has no room because it is a sacrifice you must make to contribute to the fulfilment in life. If Knecht does not give directions to Hahn, the journey cannot continue to the point of enjoyment as they will not know how to go. Hahn will be stranded, and so will Knecht. In the same way, we must be ready to make sacrifices to give direction by making contribution. Making the sacrifice is the necessary interruption that we owe one another. In teamwork, you must interrupt yourself, if not it is not teamwork. You must pay attention and be focused on others in the team if not is not a team and certainly not teamwork.

The statement by Hahn brings it all home to roust when he said that making Knecht smile gave him purpose. That should be the mantra of every human being to make another human being smile. It is a responsibility and duty we owe one another. Despite his defect, he felt that he had something to contribute to make humanity.

To make their trail a purposeful one, Knecht had to leave her wheelchair behind. It sounds metaphorical, and it appears to be because her wheelchair is equivalent to leaving her independence and dependence behind to embrace interdependence. Can you imagine how inconveniencing it would have been if she had carried her wheelchair along on that hiking? Can you imagine if Hahn had carried his walking cane or the seeing eye dog on that trail along with carrying Knecht? It would have been an impossible task. The thing is that both the dependence and independence that they each enjoyed on their own were now encompassed in interdependence thereby giving way to purposeful living. Just like with the dream team when they were together able to accomplish what was unimaginable, so also with interdependence, you can accomplish what independence and dependence can never bring you to accomplish.

From the story of these two persons, there are some vital life lessons that we can draw to teach us the significance of interdependence.

That,

- as human beings, we have defects that we cannot correct by ourselves individually.

- We also learn that those defects are not meant to inhibit our resolve in the world and this life.

- We have also learned that the unique- ness of each one of us is very critical to our existence, and we should not resign to such conditions and situations.

- We also learned that we need to express our interests regardless of the defects that we have, and that we should not just conclude that '*I cannot ...*" The reason is that until you speak out, you do not know what is in the other human being that might complement you or correct your defect.

- We have learned that there is a willing- ness, and until we get out of our self-imprisonment, we will not locate and dis- cover our freedom.

The fact is that everyone has some defects. There is no exception. One may deny such defects but that does not mean that they are absent. No one wants to see the defects or to be told, but it is there, and you need another person to correct that defect. The truth is that it is in race, status, color, age, gender, etc. The defects that you possess naturally do not make you any less of a human, neither do they make you less valuable. The one thing is that there are strengths and abilities that you possess that will complement another person's defects. This is not a case of love or marriage, but it is the real essence of one's humanity. This is the domino effect of how we inter-connect, connect, and interdepend. We must note, however, that the inter-connectedness and interdependence do not deny your independence except that your full potential is not realized without the other person or persons. Just like we saw in the dream team, they each have their independence and potentials, but they had things that would make them to exercise dependence, and they needed to interdepend for their fulfilment to become a reality in their human existence. As a human, you are defective, and full of blemishes. You are not expected to be dependent because you have a role to fulfil by addressing the defects of another person, while other address your own defects. There are some people who do not see the need to connect with others to complete them by ad- dressing their defects. Such people are full of ego and self-deception. Self-deception because they live in self-denial thinking that they are perfect and sufficient in themselves. The end-result is what we encounter many times in the society. The failure to connect always leads to suicide, homicide, depression that degenerate into mass killings. Opening to other people, in the spirit of interdependence, would lead someone to deal with the defects that would have otherwise been uncontrolled, and thereby leading to tragedy and untold pain. We need to get to the point when we will share the defects and blemishes to correct them. Communication matters. If we accept that we are created for interdependence, we will go a long way to have our defects addressed. We ought to know that our identity should not be tied to our defects.

Interdependence and Crime Against Humanity

I AM CONVINCED BEYOND human imagination that the embrace of interdependence and living in that conscience will reduce crime against humanity. Most of those committing these crimes that we bemoan everyday act independently, and there is always something missing and needing to be filled or replaced in their lives, but isolationism keeps them from sharing, hence they act alone. Those kinds of people have defects that are never addressed. Another reason is that if they value the contributions from other people, as making their own lives complete, they will know that the elimination and annihilation of such people will lead to unfulfilled life.

Interdependence and Leadership

IN THE WORLD OF INTERDEPENDENCE, the leader occupies the office of the facilitator, coordinator.

The leader is not perceived as one who knows better than the other people who are a part of the team. The leader, by title, is only for orderliness, not for being more knowledgeable or powerful. It is a functional role, not performance. The leader recognizes that he/she is not different from the members of the team. In a named team, such as a company, the leader is a staff like every other staff. Both the leader and the led are all the same in different kinds of ways, especially, in terms of functionality. Every member of the team understands that the members depend on one another for performance. Embracing this concept makes the organization to have outputs that are profitable. In an organization where the leader lords it over the people, there is always chaos resulting in non-performance. Where such exist, more like- ly than not, the leadership style has a lot to do with it. The leader probably does not coordinate or facilitate the affairs of the organization or entity very well. At a gathering early in 2022, I shared a point with a group of units staff and told the group that when anyone that is a part of their group is being accorded 'kudos', the members of the group should pat their backs, whether their names are mentioned or not, be- cause whoever is being acknowledged was able to complete that task due to the contributions of everyone in the group. That is basically the truth. It is a case of *'I perform because you performed'* or *'You performed, therefore, I per- form'*. This is the tenet of the interdependence concept. The question is: *what type of a leader are you*? This question can answer the question of: *what kind of an organization is she?* This is not to put a burden on the leader, but it must be noted that the popular saying: "*unease lies the head that wears the crown*" is begging the question of the leadership style of the lead- er. It is saying that if the leader embraces the style of leadership that promotes interdependence, then the organization will thrive. That is because the leader will recognize that the staff is the most valued asset of the organization. By recognition, it is implied that the staff is well catered for, not out of greed by the staff or the leader, but by cooperation, mutualism, and focus on the things that matter to the progress of the organization.

Do not let the title you hold get in your head because it is only a privilege that can be taken away anytime. For whatever you are doing, somebody else is training for it to take it from you, hence do not walk over other people because you will need them to function and per- form. Anybody can become who you are now, the difference is in the way you strategize for synergistical procedure for optimal output. The more you despise another person, the less you become. Think of it. It is the relatedness with another person that brings out the best in you, hence the less you relate to other people, the less your potential is displayed. The more the colors blend the more each color is projected and the more loudly each color shines. 'Relate' because you are not better than anyone, it is only a privilege. You are where you are because the organization wants to avoid chaos and embrace orderliness not because of your intelligence. When you see it that way, you will notice great changes in your outlook on life, and you will be a better person living life in joy and happiness. This approach to life is possible and plausible in the environment of interdependence because superiority is consumed in interdependence.

Chapter 6 - Interdependence: The African Perspective

Interdependence and African Culture

SAWABONA

"SAWABONA" is a common greeting among the tribes of northern Natal in South Africa, and it literally means "I SEE YOU", which means, "I respect and acknowledge you for who you are." The response to that greeting is: "SIKBO- NA" which literally means "I AM HERE" as to say, "When you see me you bring me into existence". When this happens, everyone repeats "Sawabona" which means "I respect you, value you, and you are important to me" and that per- son responds "Shikoba", which means "so... I am good and exist for you." This is the essence of humanity, that the consciousness and realization of one's existence is aligned with the consciousness of like creatures. It is the actualization of one's existence. Without this inter- action, one is left to wander who he/she truly is. There is an affirmation that is critical in this interaction with other like beings. A human will not find that actualization in animals that are non-humans. This distinguishing factor in humanity is what makes humans unique. Imagine a world in which such recognition does not exist. It can only exasperate loneliness and isolation which could result in untold consequences. This act of recognition rebuilds the damaged insides of the person by letting them know that they are loved and valued. The defects are mended. The greeting alone has the ultimate purpose of helping others and one another on their journey of self-discovery. It brings out the worth and dignity of each person.

Sawubona, is an ancient isiZulu greeting which means: *We see you.* It is equivalent to Hello and Namaste. So, when people meet and the greeting exchange is "sawubona" [we see you], the response is "yebo, sawubona" [yes, we see you too].

Why does he/she say, "*We*" see you even when it is just me, a single individual person, greet- ing you, and why do you respond by the same when it is just you, a single individual person greeting me?

In Zulu tradition, the "I" is connected to an an- cient lineage of ancestors which are always with one. So, when I meet you, not only is my- self meeting you, but my ancestors whom I am representing meet you too.

So sawubona [we see you] means me and my ancestors see you, and your response means you and your ancestors see me too.

Sawubona is more than just a greeting, it also means: We see each other, we ac- knowledge each other, we recognize each other.

Sawubona is an invitation to a deep witnessing

and presence. It invites us to communicate, to explore the possibilities of helping each other. It means we witness your journey. It spells interdependence as one whole unit.

In Business, the concept of Sawubona fosters production and productivity because as entrepreneurs the recognition is not only to greet the customers, but to also acknowledge them, to recognize (accept) them, to witness them, their feelings, their aspirations. Entrepreneurs that thrive are those that are obsessed about their customers. They invest in connecting with, and importantly, listening to their customers. They truly see their customers. This encourages patronage which leads to successful and profit- able ventures. Establishments that capture this concept do well. It is capitalism at its best because it is not ruthlessly walking over people but caring for them. It is also an investment in people which will continue to yield dividend. When an establishment invests in people, the yield will continue to flow in, but when the interest is in capital other than human, the yield will be very short lived and unrewarding.

If you do not see someone, you will not affect them and as a result you are unlikely to affect positive change to them. That is where the challenge comes with the concept of independence which emphasizes just the person; and dependence which considers the other person as more of a nuisance. Interdependence, however, acknowledges the importance of the other persons/people by acknowledging them. When I think Sawubona, I think interdependence.

Relationship problems start when people do not see, acknowledge, and recognize each other's hopes and aspirations, and contributions. In moments of conflict, seeing someone whole is both a noble and a difficult thing. It is a worthwhile pursuit because that is where possibility lives. Seeing goes beyond the physical, it is about the internal, the emotions. It is about care, about warmth, about doing work that matters. When we see people, we see beyond figures or numbers, or customers, beyond sales prospects and beyond a market gap to be exploited in business ventures. There is a connection that transcends the number mentality or statistical configuration. The person is not just mere person, but when we truly see people, the connection makes one see their hopes, dreams, desires, and the added values that might be invisible. These values become an encouragement for more search through the interaction with the persons. When we truly see people, we see people who need our humility, our undivided attention, and our assistance through our willingness to connect and share those human values embedded in us that could become added values to our own existence. We need to see each other more. This really ascertains that my humanity is fulfilled because I can relate to a like-kind creature.

The philosophy of independence is – I see my- self only.

The philosophy of dependence is - I see you only, and almost without my own existence. As a matter of fact, I do not have an existence without you because of my dependence.

The philosophy of interdependence is: We see each other/one another.

Independence and dependence, on their own, defy the true essence of humanity. The evolvement of humanity is into the discovery of the need for interconnectedness which is impossible without the interdependence concept and reality. There is no I without Us. There can be no independence without interdependence.

Ubuntu

UBUNTU IS AN AFRICAN philosophical concept of Oneness, an eternal value, based on the inter- connectedness of life. The thing that connects all life is LOVE. This concept thrives on equality of life. The concept, therefore, leads to the translation of Ubuntu as: "I am because we are".

If you are connected to one another, there is no way that one person is rising from one level to the other level that he/she will not lift the others along. It becomes very difficult to leave some people at the bottom perpetually.

Ubuntu is a philosophical concept that is based on the same foundational knowledge as phenomenology. It is based on consciousness.

Ubuntu means love, truth, peace, happiness, eternal optimism, inner goodness, etc. Ubuntu is the essence of a human being, the divine spark of goodness inherent within each being. From the beginning of time the divine principles of Ubuntu have guided African societies.

Conclusion

A Synergistic Approach to Existence

From the discourse on the interaction between dependence, independence, and interdependence, we have been able to establish that the dependency approach to living would make life and living to be burdensome; independency approach to be a myth because in the scheme of things there is really no independence for any living being; and that the interdependency approach encompasses both the dependence and independence. This is the synergetic approach to living. The interdependency approach recognizes that while human is independent, the realization of that reality resides in the dependency factor. What that means is while one is independent, he/she is equally dependent, not in the sense of being a parasite, but in the sense that the full potential becomes manifested in the dependence on other human beings. It is the interaction between the two that brings out the interdependency of the human existence.

There is a synergistic dynamic force that is created in existence which emanates from this fusion of dependence and independence. The awareness of this fact adds credence to a more fully understanding of what living life hands us. Imagine a world where everyone thinks interdependently, there would be a recognition of the significance of every human being, and the acknowledgement of the valuable contribution that such a being would make. Such awareness would create an avenue for recognition, respect, and celebration of the uniqueness of human. The unity that we so desire that have become far-fetched would be a common reality. As simplistic as it might sound, the co-existence among human would be realizable because no one would want to isolate or ostracize another for fear of destroying what would make one complete. The person you would take out might be the one who has what it takes for you to be complete. The significance of interdependence is also that it cuts across every barrier that we have built because of our quest for in- dependence – race, color, sex, religion, etc., no one can fully determine who is the carrier of what completes the other person.

Think interdependently, Act interdependently.

Independence and dependence are interdependence concealed.

Interdependence is dependence and independence revealed.

In interdependence, the strong force of both dependence and independence are revealed.

Interdependence is a package that defines hu- man existence and brings out the real meaning of life.

REFERENCES

Broodryk, J. 2006. Ubuntu African life cop- ing skills: theory and practice, vol.1, no.2, pp.3-

6. Department of Education. Education White Paper 6: (2011), Special Needs: Education, Building an inclusive education and training system.

Catching Kim, Qualities of an Independent Person. September 8, 2019 - https://catchingkim.com ›Blog

Corwin, Gary, 2005, *Church Planting 101.* Evangelical Missions Quarterly: April 2005.

Cummings, David, 1987, *Programmed for Failure – Mission Candidates at Risk*. Evan- gelical Missions Quarterly: July 1987.

Dembling, Sophia, 2012, The Introvert's Way: Living a Quiet Life in a Noisy World. California: Perigee Book.

Free Dictionary

Kim, Uichol, **Guoshu Yang, Kwang-kuo Hwang**, 2006, *Indigenous and Cultural Psy- chology: Understanding People in Context.* New York: Springer Publications.

Letseka,M. (2011). Educating for Ubuntu: Open Journal of Philosophy [online]:16 March 2015.

Lipset, S. M., 1999, *The Origins of American Individualism*. Canadian Journal of Sociology 23, 4 (1999): 511-533.

MERRIAM-WEBSTER DICTIONARY

Murphy, Helen, 2019, https://people.com/hu-[1] man-interest/melanie-knecht-trevor-hahn-hik-[2] ing-adventures/[3] published on July 12, 2019 at 03:52 pm.

Nevius, John L., 1886, *Planting and Development of Missionary Churches*. Phillipsburg, NJ: Presbyterian and Reformed Publishing Company.

Orson Scott Card, 1996). Children of the Mind. New York: Thriftbooks

Power, M.D., G. Allen, 2014, *Dementia Be- yond Disease: Enhancing Well-Being*. Towson, MD: Health Professions Press.

Randburg Msila, V. 2008, Ubuntu and School Leadership, Journal of Education 44 Young,

I.M. (1990) Justice and the Politics of Differ- ence. Princeton, NJ: Princeton University press

Svensson, Frances, 2007, *Liberal Democracy and Group Rights: The Legacy of Individualism and Its Impact on American Indian Tribes.* In the Journal Political Studies, Volume 27 Issue[4] 3[5], pages 421 – 439, University of Michigan.

1. https://people.com/human-interest/melanie-knecht-trevor-hahn-hiking-adventures/
2. https://people.com/human-interest/melanie-knecht-trevor-hahn-hiking-adventures/
3. https://people.com/human-interest/melanie-knecht-trevor-hahn-hiking-adventures/
4. http://www3.interscience.wiley.com/journal/119602001/issue
5. http://www3.interscience.wiley.com/journal/119602001/issue

Taber, Charles R., 1991, The *World Is Too Much With Us: Culture in Modern Protestant Missions*. Macon, GA: Mercer University Press.

The Walker and Avant Concept Analysis (1994). Warren, MI: Springer Publishing Com- pany.

Ubuntu: The African Dream in Management, Knowledge Resources.

Woolf, Henry Bosley, Editor, 2002, Webster's New Collegiate Dictionary. Springfield, MA: G.

& C. Merriam Company.

Yusef, W. 2014. African Philosophy of Edu-

cation Reconsidered: On being human. Rout- ledge publishers, Abingdon, USA